5th International Conference on Energy Engineering and Smart Materials (ICEESM-2020)

IOP Conference Series: Materials Science and Engineering
Volume 876

Barcelona, Spain
15 - 17 April 2020

ISBN: 978-1-7138-2689-7
ISSN: 1755-8981

Printed from e-media with permission by:

Curran Associates, Inc.
57 Morehouse Lane
Red Hook, NY 12571

Some format issues inherent in the e-media version may also appear in this print version.

Institute of Physics
Dirac House, Temple Back
Bristol BS1 6BE UK

Phone: 44 1 17 929 7481
Fax: 44 1 17 920 0979

techtracking@iop.org

Additional copies of this publication are available from:

Curran Associates, Inc.
57 Morehouse Lane
Red Hook, NY 12571 USA
Phone: 845-758-0400
Fax: 845-758-2633
Email: curran@proceedings.com
Web: www.proceedings.com

TABLE OF CONTENTS

IOP Conf. Series: Materials Science and Engineering **876** (2020) 011001 doi:10.1088/1757-899X/876/1/011001

Preface

The 5th International Conference on Energy Engineering and Smart Materials (ICEESM-2020) was held during April 15-17, 2020. This year, our annual Conference was planned to be held in beautiful and sunny Barcelona, but, unfortunately, due to the spread of the COVID-19 virus around the world, we have to curry on our presentations online. The conference programme included the invited and oral presentations from scientists working in similar areas to establish platforms for collaborative research projects. The time for each invited speaker was 45 minutes, while for authors was 15 minutes each. All keynote, parallel sessions, discussions and other activities were offered in a streaming online. The conference covered the recent trends and achievements in the fields of energy engineering, environmentally friendly technologies and smart nanomaterials.

This year we have around twenty participants from Ireland, Spain, Russia, Bulgaria, Romania, China, Jordan, India, Brazil and other countries, whom we wish fruitful discussions. ICEESM2020 created a customized online platform that was a virtual environment where the attendees could present the papers and participate in all the sessions that were organized in a full immersive experience.

We hope that you and your families are keeping well and safe. We would also like to thank all of you for making this year ICEESM Conference possible, despite this very difficult time for all of us.

Conference Chair
Prof. Tatiana Perova, Trinity College Dublin,
The University of Dublin, Ireland

IOP Conf. Series: Materials Science and Engineering **876** (2020) 011002 doi:10.1088/1757-899X/876/1/011002

Committee

Conference Committee Chairs
Jordi Llorca, Barcelona Research Center in Multiscale Science and Engineering, Spain
Tatiana Perova, Trinity College Dublin, Ireland

Program Committee Chairs
Joan Josep Roa, Universitat Politècnica de Catalunya · BarcelonaTech (UPC), Barcelona, Spain
Ander Elkoro, Mondragon University, Spain
Silvia Giordani – Dublin City University, School of Chemistry, Ireland

Local Committee Chair
Lluís Soler, Universitat Politècnica de Catalunya · BarcelonaTech (UPC), Barcelona, Spain

Poster Session Chair
Olga Vladimirovna Morozova, Federal Research Clinical Center of Physical-Chemical Medicine of the Federal Medical and Biological Agency of Russia, Russian Federation

International Technical Committees
Hind Ahmed, Trinity College Dublin, Ireland
Hoda Akbari, Trinity College Dublin, Ireland
Subhash Chandra, Trinity College Dublin, Ireland
Daniel Kehoe, Trinity College Dublin, Ireland
Jaroslaw Jakubowicz, Poznan University of Technology, Poland
Zhe Xu, Dartmouth College, USA
Bolian Chen, DEEPLEARNING.AI, Taiwan
Bojan A. Marinkovic, Pontifical Catholic, University of Rio de Janeiro, Brazil
Artur Rydosz, AGH University of Science and Technology, Poland
Mohd Rafie Johan, University of Malaya, Malaysia
Ipek Guler, Cankaya University, Turkey

IOP Conf. Series: Materials Science and Engineering **876** (2020) 011003 doi:10.1088/1757-899X/876/1/011003

Peer review statement

All papers published in this volume of *IOP Conference Series: Materials Science and Engineering* have been peer reviewed through processes administered by the proceedings Editors. Reviews were conducted by expert referees to the professional and scientific standards expected of a proceedings journal published by IOP Publishing.

IOP Conf. Series: Materials Science and Engineering **876** (2020) 012001 doi:10.1088/1757-899X/876/1/012001

Intracellular delivery of nanoparticles fabricated from monoclonal and polyclonal antibodies against influenza A and hepatitis B virus antigens

O.V. Morozova [1, 2, †], E.R. Pavlova [1,3], E.I. Isaeva [2], A.I. Sokolova [1, 4], E.A. Obraztsova [1, 5], E.A. Ivleva [1] and D.V. Klinov [1,3]

[1] Federal Research and Clinical Center of Physical-Chemical Medicine of Federal Medical Biological Agency
[2] Ivanovsky Institute of Virology of the National Research Center of Epidemiology and Microbiology of N.F. Gamaleya of the Russian Ministry of Health
[3] Moscow Institute of Physics and Technology
[4] Moscow State University, Faculty of Biology, Department of Bioengineering
[5] M.M. Shemyakin and Yu.A. Ovchinnikov Institute of Bioorganic Chemistry of the Russian Academy of Sciences

[†]Corresponding author's e-mail address: omorozova2010@gmail.com

Abstract. Specific antibodies (AB) may serve as molecular tools for both targeted, etiotropic and immunomodulation therapy of the infectious and oncological diseases. Despite their ability to bind with extracellular virions and isolated viral antigens AB may not pass through host cell membranes to inhibit intracellular infectious agents. Nanoparticles (NP) were fabricated from polyclonal and monoclonal AB against the influenza A structural and non-structural proteins as well as against the hepatitis B virus surface antigen (HBsAg). ELISA and immunofluorescent analysis revealed the ligand-binding activity of the AB NP comparable with the properties of the original AB. Fluorescent microscopy and quantitation showed efficient cellular uptake of the labeled AB NP and their degradation in 5 days posttreatment. Intracellular AB NP induced interferon (IFN) gene expression but not interleukin (IL) 4 and IL10 transcription. The cytokines mediated immunomodulation potential of the protein NP.

1. Introduction

Specific AB are used for targeted immunotherapy. Among 75 approved and commercially available drugs on the base of AB more than half are checkpoint inhibitors against surface receptors. In spite of evident advantages of AB such as their high specificity, affinity and prolonged action the remaining problem is their lacking penetration in all eukaryotic cells excepting B lymphocytes, follicular dendritic cells, natural killer cells, macrophages, neutrophils, eosinophils, basophils, human platelets, and mast cells with receptors to F_c fragments of heavy chains of AB [1]. Since AB cannot entry into host cells and inhibit or neutralize intracellular viruses there then the AB-mediated treatment may not prevent a persistent infection inside cells. To eliminate viruses the specific and active AB must be delivered inside infected cells near virus replication sites. Targeted drug delivery may be mediated by AB on NP surface [2]. Since 1990 US Food and Drug Administration approved 51 nanomedicines [3]. The number of nanotherapies remains low partly due to acute and chronic toxicity. Various combination of immunomodulating cytokines, chemokines and Toll-like receptor agonists targeting various immune cells as well as immunocytokines that combine AB with cytokines have been designed, developed and evaluated in many preclinical trials [4]. Moreover, beyond providing delivery platforms different nanosystems possess intrinsic immunomodulatory properties causing pleiotropic and possible side effects [4]. Despite clinical implementation of protein NP little is currently known about their safety, cellular uptake, intracellular distribution patterns, and the clearance mechanisms [2-4]. Our goal was the fabrication of NP from polyclonal and monoclonal AB against the influenza A structural and non-structural proteins as well as against HBsAg, comparative immunoassays of their binding with specific antigens and analysis of intracellular delivery of AB NP.

2. Methods

2.1. Tissue cultures.
Human larynx carcinoma HEp-2 cells were obtained from the National Research Center of Epidemiology and Microbiology of N.F. Gamaleya of the Russian Ministry of Health, Moscow, Russia and grown in culture medium 199 with 8% fetal bovine serum in the presence of 100 U/ml penicillin and 100 U/ml streptomycin until monolayer formation.

2.2. Fluorescent labeling of AB.
The immunoglobulins (Ig) were labeled with rhodamine B dye (Rho B) in 0.1 M Na_2CO_3 solution at pH 9.3 for 1 h at room temperature or overnight in refrigerator and then purified by Sephadex G25 gel-chromatography with centrifugation at 700 g.

2.3. AB NP fabrication.
NP were constructed from AB solutions in hexafluoro-propan-2-ol (HFIP) by nanoprecipitation as previously described [5].

2.4. Circular dichroism (CD) spectroscopy.
CD spectra were obtained by Chirascan spectrophotometer (Applied Photophysics, UK).

2.5. Fluorescent microscopy and quantitation.
The HEp-2 cells with fluorescent AB NP were imaged using the fluorescent microscope Eclipse Ti (Nikon, Japan) with a Plan Fluor 40x/0.6 objective and filters providing excitation/emission for RhoB (528-553 nm/590-650 nm) and Hoechst 33342 (340-380 nm/435-485 nm). The images were captured with the camera ORCA-Flash4.0 (Hamamatsu, Japan) with the exposure time of 3-5 s. Fluorescence intensity at 580 nm of cell lyzates at different time points was measured using spectrofluorometer "Fluoromax+" (Horiba Scientific, Japan).

2.6. Cytotoxicity assay.
MTT assay was performed to evaluate the *in vitro* cytotoxicity of RhoB, fluorescent AB and corresponding NP. The absorbance was measured at 570 nm.

2.7. Cytokine gene expression.
Cytokine RNA including IFN α, β, γ, λ, IL 2, 4, 6, 10, 12, 23 were detected by reverse transcription with quantitative real time PCR (RT^2-qPCR) as previously described [6].

3. Results and Discussion
Nanoprecipitation of polyclonal and monoclonal AB of different origin, classes and specificity from their solutions in fluoroalcohol HFIP permitted to fabricate water-insoluble and stable NP without cross-linking. Mixed protein NP could be constructed by the nanoprecipitation of core albumin and its subsequent cross-linking with surface AB shells using 0.1% glutaraldehyde. The AB NP might be also constructed from blood sera containing the specific AB diluted with HFIP (ratio 1:3). Similar CD spectra of Ig in water solutions and in NP revealed AB stable secondary structures after nanoprecipitation. Noteworthy that the protein NP including those from different AB are very stable up to 1 year of storage in water at $+4^0$ C [5], in culture media and in biological fluids (blood sera,

saliva – for 5 days at 37^0 C). Binding of the AB NP with specific influenza A virus antigens and HBsAg was shown by ELISA, immunofluorescent analysis (IFA), hemagglutination inhibition (HI) and neutralization tests. According to MTT assay both AB and the AB NP with concentrations up to 2 mg/ml were not toxic for human cell lines (Fig. 1). Main part (50-80%) of the fluorescent AB NP appeared to be associated with cellular monolayers. According to our estimations each AB NP of average diameter near 100 nm contained approximately 10^4 Ig molecules in their native conformations with specific ligand-binding properties and up to 10^5 NP can be found per cell. The average size of one eukaryotic cell is known to be 10 μm, consequently, the cellular surface area $S=4\pi R^2$ is ~314 $\mu m^2=3.14*10^8$ nm^2. Thus, 10^5 NP of average diameter 100 nm may cover up to 10% of the cellular surface. However, cellular uptake of the AB NP was not simultaneous. The AB NP could be found in cells in 2 hours posttreatment with maximum after 2 days and subsequent gradual decline until background values (Fig. 2). The slow enhancement of the cell-associated fluorescent Ig NP amounts could be resulted from gradual entry of the NP into cells and further stepwise degradation.

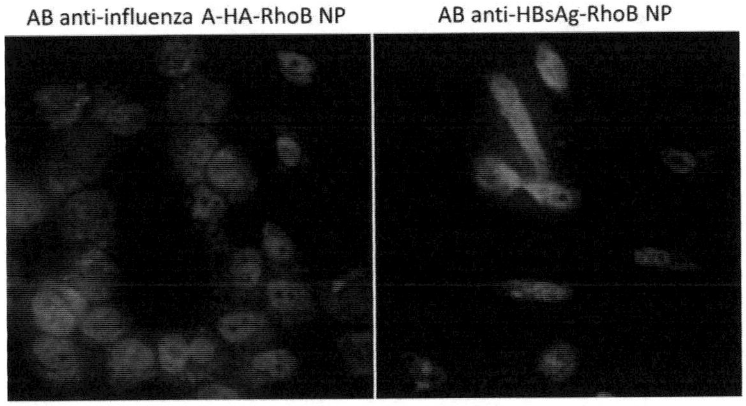

Figure 1. Cellular uptake of AB NP in 2 days posttreatment. Nuclei were stained with Hoechst 33342 (blue color), AB NP were labelled with Rho B (red color).

Cellular uptake relies on 4 known mechanisms: clathrin- and caveolar-mediated endocytosis, phagocytosis, macropinocytosis, and pinocytosis [2]. Specific clathrin- and caveolin-mediated endocytosis pathways can be excluded from consideration because of the absence of specific receptors for AB on HEp-2 cells [1]. Therefore, the fluorescent AB attachment to human cell monolayers was not essential (Fig. 2). Phagocytosis is used by macrophages, dendritic cells and neutrophils engulfing particles larger than 0.5 μm only [2]. Consequently, macropinocytosis remains the most probable pathway by which cells internalize AB NP, and large vesicles (0.2–5 μm) are formed inside cells [2]. For Rho B dye the pinocytosis seems to be a possible endocytic process by which the eukaryotic cells absorb extracellular fluids and small molecules (Fig. 2).

Because of the AB NP degradation in the eukaryotic cells (Fig. 2) their binding with intracellular infectious agents could be possible near outer cellular membranes and in endosomes during the first two days postinfections. The step-by-step cleavage of Ig on the NP surface did not exclude the antigen-binding properties of the AB from underlying, deeper and core layers. The evident advantages of the AB NP disintegration include their low toxicity and the cytokine gene expression induction.

The slow gradual increase in intracellular fluorescent NP accumulation from 30 minutes to 48 hours posttreatment and subsequent decrease for 3 days until background values suggested that the AB NP entry in the cells was not synchronous.

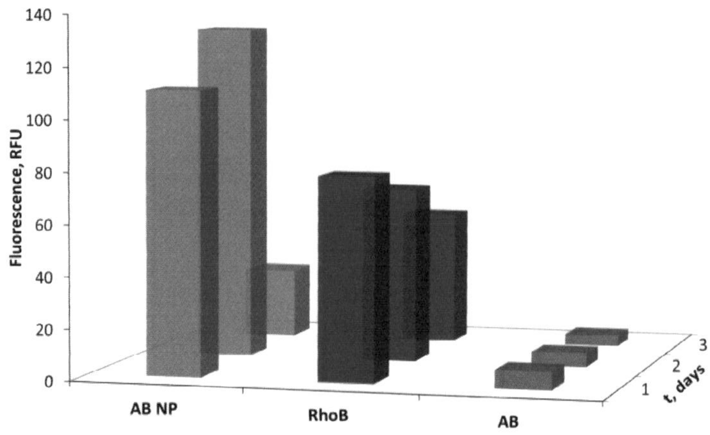

Figure 2. Comparison of cellular uptake of RhoB dye, AB and corresponding NP during 3 days after addition.

Despite immunological tolerance of cells to Ig from growth culture media and degradation of fluorescent protein NP during 3 days (Fig. 2), the AB NP induced transcription of IFN genes of types I and III in 1-2 days posttreatment. IL 4 and IL10 RNA were not detected using RT^2-PCR. The cytokine spectra and the polarization indexes (PI) calculated as ratio Th2:Th1 cytokines were similar for NP constructed from various AB. PI near 0 with the absence of Th2 cytokines suggested Th1 polarization of innate immunity induced by the AB NP. The protein NP remain non-immunogenic [7]. The absence of detectable antibodies after the protein NP administration could be caused by Th1 polarized mainly cellular immune response. The Th1 innate immunity may protect cells from invasions.
Degradation of the fluorescent AB NP in 3-5 days posttreatment together with the IFN α, β and λ gene expression could be resulted from the fluorescent AB proteolitic hydrolysis in lysosomes but not in intracellular vesicles without proteases. So, the macropinocytosis was not the only possible way of the AB NP endocytosis [2]. Exocytosis of the RhoB-AB NP and their gradual accumulation in the culture medium after 2 days posttreatment were not observed.
Taken together, the stable AB NP with the antigen-binding activity can be used for targeted delivery and immunimodulation towards Th1 mainly cellular immune response. Their intracellular binding with the virions may be limited because of proteolytic degradation after 2 days.

4. Summary
AB NP retained the ability of the original AB to bind with specific antigens, were not toxic for human cells, gradually accumulated in cells and induced Th1 polarized cytokine gene expression.

5. References
[1] Thomas D.B. Viruses and the cellular immune response. New York, Basel, Hong Kong; 1993.
[2] Oh N., Park J.H. Endocytosis and exocytosis of nanoparticles in mammalian cells. International Journal of Nanomedicine 2014:9 (Suppl 1) 51–63. DOI: 10.2147/IJN.S26592.
[3] Bobo D., Robinson K.J., Islam J., Thurecht K.J., Corrie S.R. Nanoparticle-based medicines: a review of FDA-approved materials and clinical trials to date. Pharm Res. 2016;33(10):2373–2387. DOI: 10.1007/s11095-016-1958-5.
[4] Liu Z., Jiang W., Nam J., James J. Moon, Kim B.Y.S. Immunomodulating Nanomedicine for Cancer Therapy. Nano Lett. 2018, 18, 6655–6659. DOI: 10.1021/acs.nanolett.8b02340.
[5] Morozova O.V., Pavlova E.R., Bagrov D. V., Barinov N.A., Prusakov K.A., Isaeva E.I., Podgorsky V.V., Basmanov D. V., Klinov D.V. (2018) Protein nanoparticles with ligand-binding and enzymatic activities. International Journal of Nanomedicine 2018:13; 6637–6646, DOI: 10.2147/IJN.S177627.

[6] Morozova O.V., Ospelnikova T.P. Patent 2016131205, 28.07.2016. Ru 2 627 179. C12Q 1/68 (2006.01).
[7] Podaralla S.K., Kaushik R.S., Perumal O.P. Method of Forming Non-Immunogenic Hydrophobic Protein Nanoparticles, and Uses Therefor Patent US9616021.

Acknowledgements
Our research was supported by the Russian science foundation (grant № 17-75-30064).

IOP Conf. Series: Materials Science and Engineering **876** (2020) 012002 doi:10.1088/1757-899X/876/1/012002

Shell structure of the porous TiNi-framework obtained by the SHS method

S Gunther, E Marchenko, G Baigonakova, Y Yasenchuk

National Research Tomsk State University, Russia

Corresponding author's e-mail: gat27@mail.ru

Abstract. Porous SHS-TiNi alloys, presented as a porous framework with a complex structure of the Core-Shell type, were studied by microscopic and structural methods. The core is the porous framework of an intermetallic TiNi phase. The shell was found to be a complex three-layer complex. An oxygen-hardened peritectic TiNi–Ti_2Ni layer 1-6 μm thick forms an inner shell layer. Thermo and corrosion-resistant amorphous-nanocrystalline intermetallic oxycarbonitrides, saturated with O, N, C interstitial impurities, form middle and outer layers. The middle dense double layer 60-80 nm thick protects the TiNi matrix. The outer layer 1-15 μm thick has a loose layered structure, it is enriched in the MAX-phases Ti_3SiC_2, Ti_3AlC_2. The dense nano double layer provides a chemical passivity to the porous nickelide titanium alloy with shape memory, does not limit mobility and improves the ability to integrate into biological tissues.

1. Introduction

Porous TiNi-based intermetallic alloys obtained by self-propagating high temperature synthesis (SHS) are used in clinical practice. They possess biocompatibility, which is confirmed experimentally [1,2]. The phase composition and structure of the surface layers completely determine the morphological features of the surface and the biochemical compatibility of implants with biological tissues.

The morphology, phase composition and structure of porous alloys of the Ti-Ni system have been studied in many scientific works, which show that these alloys almost always contain inclusions of intermetallic oxycarbonitrides $Ti_4Ni_2(O,N,C)$, as well as titanium oxides, carbides, and nitrides of various stoichiometry [3-6]. Their presence is due to the high cost of purifying reagents and reaction products from interstitial impurities [7].

The SHS method has features that allow the creating of porous alloys with increased corrosion resistance, compared with the reaction sintering method of Ti-Ni system alloys [8,9]. They successfully resist the corrosive effects of a chlorine-containing medium due to the surface shell. However, the phase composition and structure of the surface layers of porous SHS-TiNi alloys have not been studied sufficiently, therefore, studies on this solve remain relevant.

2. Material and methods

Self-propagating synthesis reaction of the TiNi alloy was carried out through layer-by-layer regime. We used titanium powder reduced with RP-Ti grade OM-2 by POLEMA (Russia) and Carbonyl Nickel Powder PNK-1L5 by NORILSK NICKEL (Russia). The porous alloy was disintegrated into granules by impact destruction. The granule filling was annealed in a vacuum at a temperature of 1200 ° C for 40 min. An annealing was carried out in order to diffuse the granules sintering.

The structural-phase composition of the surface was studied by XRD analysis using an XRD-6000 diffractometer with Cu Kα radiation. The POWDER CELL 2.4 full-profile analysis program was used to process the diffraction pattern. Optical micrographs were obtained by means of Carl Zeiss Axiovert 40 MAT light microscope and the Focus stacking technique. Confocal laser scanning microscopy was performed on a surface area of 30×30 mm of a porous sample using irradiation lasers with characteristic excitation wavelengths of 405 and 488 nm and ChS1 filters: 517-695 and 410-695. Electron microscopic studies were carried out using a JEM-2100F microscope (JEOL). The lamella for TEM was prepared on QUANTA 200 3D. Plates were cut 0.7 mm thick, 7 mm wide and 80 mm long from porous SHS-TiNi samples. The plates were destroyed on an Instron installation by tensile strain. Fractogram analysis of fracture surfaces was carried out using a QUANTA 200 3D scanning electron microscope in SEM mode.

3. Results and discussion

Confocal laser scanning microscopy (CLSM) of polished porous SHS-TiNi samples shows a presence of the shell (green color) on the porous framework, marked C (Fig.1, inset 1). The shell of the porous framework is also clearly visible due to its brightness on the optical microscopy image in the dark field, in contrast to the dark cross-section surface (Figure1, inset 2).

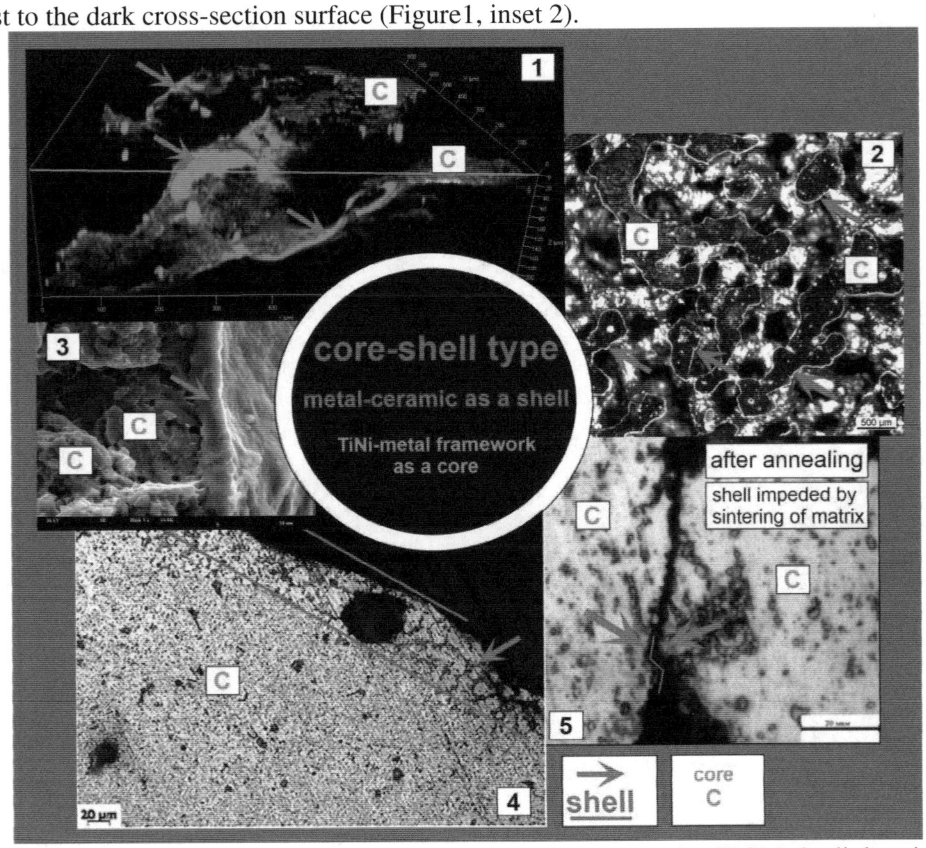

Figure 1. Scheme of shell view by various types of microscopy: 1 - CLSM; 2 - light microscope cross section in a dark field; 3 - SEM after impact failure; 4, 5 - light microscope cross section in a bright field; 2, 4 - before annealing; 5 - after annealing

The shell is a feature of the SHS process of intermetallic compounds, and it is formed during the interaction of reaction gases with the melt film on the surface of the porous framework [10]. Due to the heat of the reaction gases along the walls of large pores, in the zones of peritectic crystallization, the internal structure of the framework is enlarged: the intergranular peritectic Ti_2Ni phase and TiNi grains (Figure 1, inset 4). The Ti_2Ni phase is more saturated with oxygen and hardens than other phases in the Ti-Ni system. The TiNi phase less dissolves oxygen, but partially loses titanium. A change in composition towards nickel enrichment induces it to go into a martensitic state. This leads to the hardening of TiNi grains. Thus, a hardened layer is formed, which is destroyed by a brittle mechanism under shock loads (Figure 1, inset 3). In contrast, a non-hardened matrix of the framework is destroyed by a viscous mechanism.

The thick of the hardened shell is determined by the depth of oxygen diffusion and reaches 1-6 μm. The surface layer prevents sintering the granules of the porous framework in a vacuum (Figure 1, inset 5). At an annealing temperature above 1200 °C, the TiNi phase becomes viscous and easily deforms under the influence of surface tension forces. However, the dense contact of granules does not lead to the formation of diffusion necks between them, even when holding for 40 minutes. The presence of volume diffusion and the absence of surface diffusion indicate that the surface layer is refractory. Previously, the annealing of TiNi powder under the same conditions led to the formation of diffusion necks between the particles and a complete change in their morphology as a result of surface and bulk diffusion.

It was previously established that the surface of the porous SHS-TiNi is corrosion resistant [11]. The surface part of the shell was detected at TEM in the form of a double layer 60-80 nm thick (Figure 2). Chemisorption of dissociated reaction gases by a titanium-enriched melt occurs during peritectic crystallization at the beginning of the structuring zone. Therefore, we found the double layer even in alloys with small pores in the range of 50-70 μm, which did not have time to grow larger.

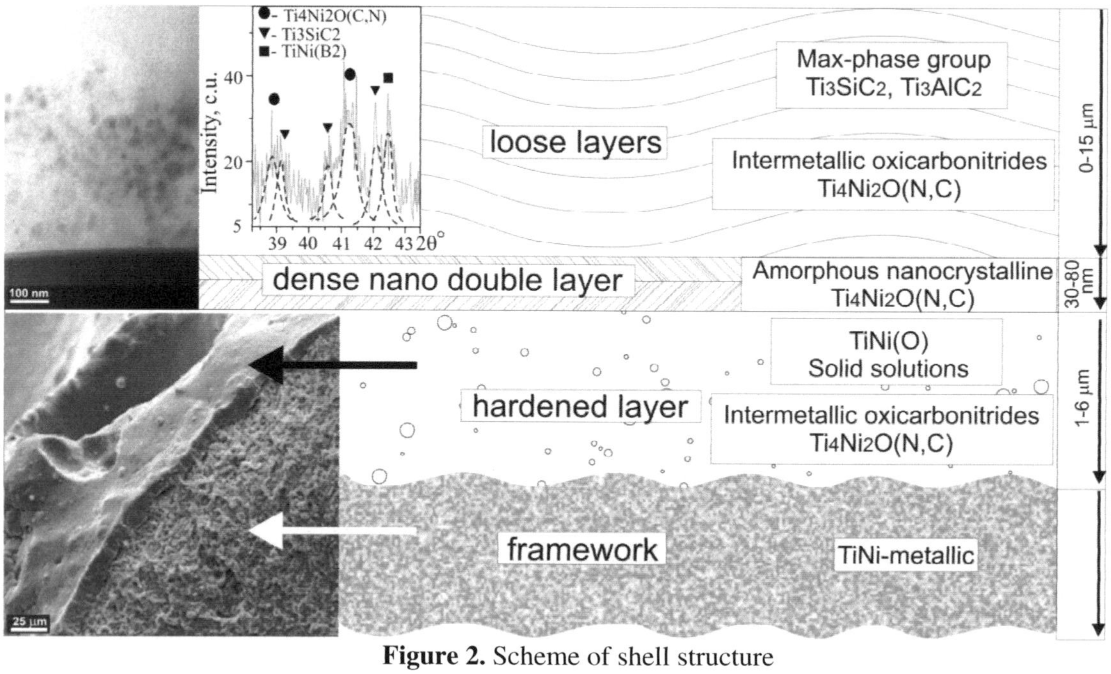

Figure 2. Scheme of shell structure

Chemoepitaxy forms the double carbonitride film on the liquid surface of the melt in two stages. At the first stage, as a result of chemisorption of O, N, C, Si, Al impurities from the reaction gases, an amorphous Ti-Ni-C-N-O-Si-Al film is formed on the peritectic liquid surface. At the second stage, as a result of the crystallization of the amorphous Ti-Ni-C-N-O-Si-Al film, the epitaxial growth of nanocrystalline phases $Ti_4Ni_2O(N,C)$, Ti_4Ni_2O, Ti_3SiC_2, Ti_3AlC_2 occurs.

Considering the capability of the Ti_2Ni intermetallic compound to amorphize under the influence of interstitial impurities, the amorphous phase of intermetallic oxycarbonitrides is formed after the passing the synthesis wave. At the structuring stage, the amorphous phase crystallizes under the thermal influence of the reaction gases and turns into amorphous nanocrystalline. It is known that such amorphous – nanocrystalline phases have high strength and corrosion resistance [12-14].

A liquid layer of the peritectic shell of TiNi grains, whose composition is close to the stoichiometry of the Ti_2Ni, remains under surface amorphous film after its hardening. This allows the framework, which consists of viscous TiNi grains, liquid interlayers, a solid, but thin shell, to remain in a mobile state throughout the entire structuring stage of the porous alloy. In this case, under the influence of the reaction gases heat, the pores merge and enlarge, as well as framework walls thicken.

The reaction gases, flowing out of the reaction zone, capture part of the melt and carry compounds of impurities with titanium: Ti_4Ni_2O, Ti_3SiC_2, Ti_3AlC_2 in the form of an aerosol over the structuring zone. Thus, loose layers are formed on the surface of the double dense layer.

Qualitative XRD analysis has shown that the main component of the surface layer is the amorphous-nanocrystalline phase of intermetallic oxycarbonitrides $Ti_4Ni_2(O,N,C)$, which is a solid solution of O, N, C in the peritectic Ti_2Ni phase. Diffraction reflections from cermet ternary MAX phase Ti_3SiC_2 with a nanolaminate structure and the intermetallic matrix TiNi B2 were revealed in the diffractograms.

Figure 3 shows a TEM image of surface layers with corresponding SAED patterns. The Ti_4Ni_2O phase with residual amorph was detected in all layers. Diffraction patterns *1* and *2*, taken from loose and dense nano double layers, show a diffuse halo near small diffuse reflections of the Ti_4Ni_2O phase, which indicates an amorphous-nanocrystalline state. The previously obtained XRD spectrum from the surface of the SHS-TiNi shell contains an amorphous halo and diffuse X-ray reflections of Ti_4Ni_2O with CSR=7 nm, which confirms its amorphous-nanocrystalline structure [15]. Particles of the Ti_4Ni_2O phase of globular shapes were found in the hardened layer, their corresponding diffraction pattern *3* is characteristic of a fully crystalline structure.

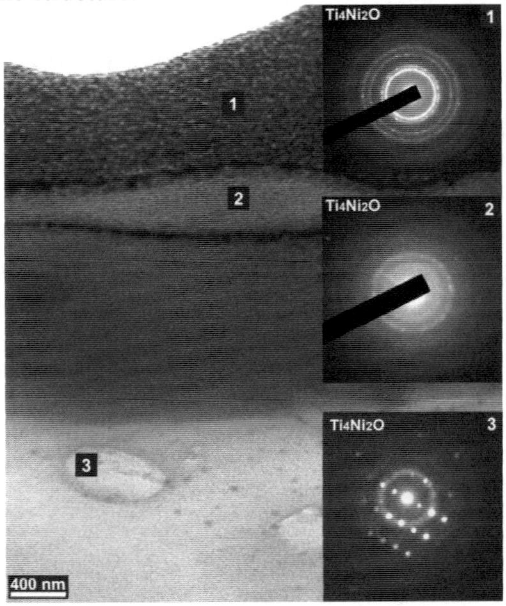

Figure 3. The TEM image of the shell with corresponding SAED patterns

4. Conclusions

1. It was experimentally shown that, as a result of the interaction of reaction gases with the peritectic melt in the SHS process, on the porous TiNi-framework the shell is formed. The shell consists of the inner hardened layer 1-6 μm thick and the outer corrosion-resistant nanocrystalline layer 15 nm-15 μm thick.
2. The corrosion-resistant nanocrystalline layer was shown to also consist of the inner dense double layer with a thickness of 15-60 nm and the outer loose layer 0-15 μm thick. The dense inner layer is formed by intermetallic oxycarbonitrides $Ti_4Ni_2(O,N,C)$. The outer layer is formed by oxycarbonitrides Ti_4Ni_2O and MAX-phases Ti_3SiC_2, Ti_3AlC_2.
3. Due to the thermal and corrosion resistance, the dense double nanocrystalline layer provides reliable protection of the porous TiNi-framework both during electrochemical and gas corrosion. The small thick of the inner double layer and the friability of the thicker outer layer do not limit the mobility of the superelastic framework. The loose multi-layer structure of MAX-phases provides good integration into biological tissues.

IOP Conf. Series: Materials Science and Engineering **876** (2020) 012002 doi:10.1088/1757-899X/876/1/012002

5. References

[1] Gunther V E 2000 *Delay Law and New Class of Materials and Implants in Medicine* (Northampton, MA: STT) *pp*
[2] Yasenchuk Y, Marchenko E, Gunther V, Radkevich A, Gunther S, Baigonakova G, Hodorenko V, Chekalkin T, Kang J H, Weiss S and Obrosov A 2019 *Materials* **12** 2405–25
[3] Aydogmus T and Bor A S 2011 *Turkish J. Eng. Env. Sci.* **35** 69
[4] Lelatko J, Freitag M, Rak J, Wierzchon and T, Goryczka T 2012 *Solid State Phenom.* **186** 259
[5] Li B Y, Rong L J, Li Y Y and Gjunter V E 2000 *Acta Materialia* **48** 3895
[6] Borovinskaya I P, Gromov A A, Levashov E A, Maksimov Y M, Mukasyan A S and Rogachev A S 2017 *Concise Encyclopedia of Self-Propagating High-Temperature* (Amsterdam: Elsevier Inc.)
[7] Holt J B and Dunmead S D *Annual Review Material Science* **21** 305–34
[8] Whitney M, Corbin S F and Gorbet R B 2009 *Intermetallics* **17** 894
[9] Novák P, Kadlecová B and Salvetr P 2017 *Procedia Engineering* **184** 681
[10] Kim J S, Yang S G, Kang J H, Kang S B, Yoon K S and Kwon Y S 2004 *Mater. Sci. Forum* **449-452** 1097.
[11] Gunter V, Yasenchuk Yu, Gunther S, Marchenko E and Yuzhakov M 2019 *Mater. Sci. Forum* **970** 320.
[12] Tavoosi M and Barahimi A 2017 *Surfaces & Interfaces* **8** 103
[13] Bacha N E 2017 *3rd International Multidisciplinary Microscopy and Microanalysis Congress* (Springer International Publishing AG) pp 129-133.
[14] Liu X, Xiang Zh, Niu J and Xia K 2015 *Int. J. Electrochem. Sci.* **10** 9042
[15] Marchenko E, Yasenchuk Yu, Gunther S, Baigonakova G, Gunther V, Chekalkin T, Weiss S, Obrosov A, Dubovikov K 2019 *Mater. Res. Express* **6** 1165

Acknowledgments

This research was supported by Russian Science Foundation (grant № 18-12-00073)

IOP Conf. Series: Materials Science and Engineering **876** (2020) 012003 doi:10.1088/1757-899X/876/1/012003

Numerical modelling of PCMs encapsulated in spherical shells using inverse enthalpy-temperature functions

O G Pop[a*], C A Iuga[b,c], L. Fechete Tutunaru[d], F. Domnita[a], M C Balan[e]

[a]Department of Building services engineering, Technical University of Cluj-Napoca, Romania.
[b]Department of Pharmaceutical Analysis, Iuliu Hatieganu University of Medicine and Pharmacy, Cluj-Napoca, Romania.
[c]Iuliu Hatieganu University of Medicine and Pharmacy, Proteomics and Metabolomics Department, MedFuture Research Center for Advanced Medicine, Louis Pasteur Street 4-6, Cluj-Napoca, 400012, Romania.
[d]Department of Automotive Engineering and Transports, Technical University of Cluj-Napoca, Romania.
[e]Department of Mechanical Engineering, Technical University of Cluj-Napoca, Romania

*E-mail: Octavian.Pop@insta.utcluj.ro

Abstract. The study presents a numerical model that uses the enthalpy method in order to solve the moving boundary problem and to predict the thermal behaviour of phase change materials during solidification and melting. The commercially available paraffin RT21 is considered as phase change material, and the thermophysical properties of the paraffin were determined by differential scanning calorimetry. An inverse enthalpy-temperature function was determined based on the results of the differential scanning calorimetry measurements in order to model the phase change process. The model was experimentally validated, and good agreement was found between the computed and measured results.

1. Introduction

Phase change materials (PCMs) have the potential of reducing energy consumption due to their capacity of storing high amounts of thermal energy at relative constant temperature, thus, eliminating the gap between energy availability and peak consumption periods [1].

Mathematical modelling is an important instrument to be employed in simulating the behaviour and in evaluating the efficiency of PCM based regenerative heat exchangers [2]. In [3, 4] it is considered that mathematical modelling should be accompanied by experimental research that generates appropriate data for validation, as confidence in existing models is relatively low.

The accuracy of models is significantly influenced by: the variation of the thermophysical properties of the PCMs [5, 6], and the interior natural convection that occurs during melting [7].

The importance of the precision at which the thermophysical properties are determined when generating input data is studied in [5, 6]. Differential scanning calorimetry (DSC) is the main thermo-analytical method employed in order to determine the thermal behaviour of the PCM during phase change. Based on the thermal flux indicated by the DSC the variation of enthalpy or apparent heat capacity is determined and taken into account [8, 9].

The internal heat transfer mechanisms also play an important role in the modelling procedure. The solidification process is characterized by conduction, while melting is dominated by convection [7]. The most common method of modelling natural convection is by incorporating this heat transfer mechanism in an increased effective thermal conductivity [7].
The aim of this study is to present a mathematical model that estimates the enthalpy and temperature profile of PCMs encapsulated in spheres. The model uses inverse enthalpy-temperature functions to predict the thermal behaviour of the PCM during melting and solidification. An experiment is carried out in order to generate data and validate the mathematical model.

2. The mathematical model and enthalpy formulation

The enthalpy method was adopted in order to model the thermal behaviour of the PCM. The PCM undergoes a heat transfer process through forced convection using air as heat transfer fluid. The governing Fourier equation of the enthalpy profile within the spherical capsule is:

$$\rho_{PCM}\frac{\partial H_{PCM}(r,t)}{\partial t} = \frac{\lambda_{PCM}}{r^2}\left[\frac{\partial}{\partial r}\left(r^2\frac{\partial T_{PCM}(r,t)}{\partial r}\right)\right] \quad (1)$$

where ρ_{PCM} [kg/m^3] is the density of the PCM, H_{PCM} [J/kg] is the enthalpy of the PCM, r [m] is the radial coordinate, t [s] is the time, λ_{PCM} [W/m·K] is the thermal conductivity of the PCM and T_{PCM} is the temperature of the PCM.
The boundary conditions on the exterior surface of the sphere and at the centre are:

$$m_C c_C\frac{dT_C}{dt} = \alpha\cdot 4\pi r_{C,e}^2\cdot\left(T_{air}-T_{C,e}\right) - \lambda_{PCM}\cdot 4\pi r_i^2\cdot\frac{\partial T_{PCM}}{\partial r} \text{ at } r = r_i \text{ and } \frac{\partial T_{PCM}}{\partial r} = 0 \text{ at } r = 0 \quad (2)$$

where m_C [kg], c_C=840 [J/kg·K], T_C [°C] are the mass, the specific heat capacity, and the temperature of the capsule, respectively, α [W/m^2·K] is the coefficient of forced convection calculated as recommended in [10], $r_{C,e}$ [m] and $T_{C,e}$ [°C] are the exterior radius and temperature of the capsule, respectively, T_{air} [°C] is the air temperature and ri [m] is the inner radius of the capsule.
The explicit scheme was employed to numerically solve the governing equation, by computing the new value of the enthalpy based on the old values of enthalpy and temperature. The exterior boundary condition was solved explicitly neglecting the temperature gradient in the small thickness of the spherical wall. The number of spatial nods and the time increment were set at 7 and 0.5 s, respectively. The effective thermal conductivity of the PCM in liquid phase during melting was determined using the following correlation $\frac{\lambda_{eff}}{\lambda_{PCM}} = 0.174Ra_\delta^{0.323}$ valid for $10^3 < Ra_\delta < 5\cdot 10^8$ [7]. The Rayleigh number Ra_δ [-] is determined using δ [m], the gap between the solid front and the spherical wall, as characteristic length. A program, implementing the numerical solution, was developed in Matlab.

3. The thermophysical properties of the PCM and the enthalpy function

RT21 a paraffin wax produced by Rubitherm GmbH was considered in this study as PCM. The thermophysical properties were determined in a Mettler Toledo 822 DSC under dynamic atmosphere at an air flow rate of 25 ml/min. A 10.14 mg sample was sealed in a 40 μL aluminium crucible and scanned in the temperature range of (6-27) °C, at a cooling and heating rate of 0.1 K/min. Figures 1 and 2 present the DSC results. The PCM has the following properties: density in solid/liquid phase ρ_{PCM} =880/770 kg/m^3, the specific heat capacity c_{PCM} =2 kJ/kg·K, the thermal conductivity λ_{PCM} =0.2 W/m·K, the cinematic viscosity ν_{PCM}= 25.71·10^{-6} m^2/s, and the latent heat for solidification/melting L_{PCM} = 136.33/138.00 [kJ/kg]. The start and stop temperatures of the PCM for the solidification process are 23.44 °C and 6.00 °C, respectively. For the melting process the start and stop temperatures are 6.66 °C and 24.13 °C, respectively. At these temperatures the DSC curve separates from the baseline. The enthalpies corresponding to the start/stop temperatures for the solidification process are 181.818/12.228 kJ/kg and for the melting process are 13.445/179.582 kJ/kg. These enthalpies also represent the validity interval of the inverse temperature-enthalpy function.

IOP Conf. Series: Materials Science and Engineering **876** (2020) 012003 doi:10.1088/1757-899X/876/1/012003

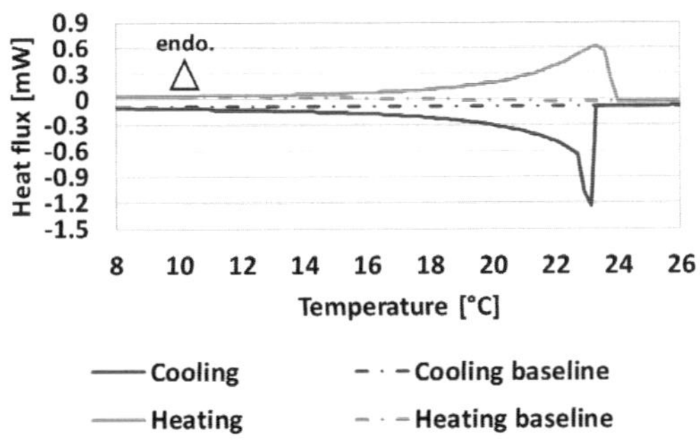

Figure 1. The heat flux variation

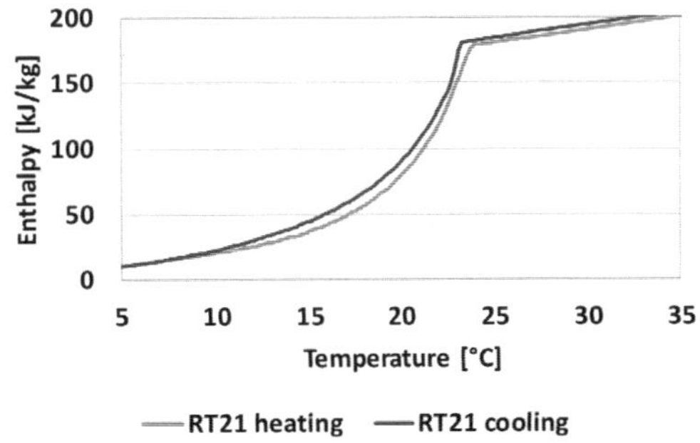

Figure 2. The enthalpy variation

In order to compute the PCM temperature based on the resulting enthalpy variation, an inverse exponential enthalpy-temperature function was considered. For example, the inverse enthalpy-temperature function for melting is as follows:

$$T_{PCM} = \begin{cases} \frac{H_{PCM}}{c_{PCM}} \; for \; T_{PCM} \leq T_{s,m} \\ y_0 + A_1 \cdot \exp\left(\frac{H_{PCM}}{C_1}\right) + A_2 \cdot \exp\left(\frac{H_{PCM}}{C_2}\right) for \; T_{s,m} < T_{PCM} \leq T_{st,m} \\ T_{st,m} + \frac{H_{PCM} - c_{PCM} T_{s,m} - L_{PCM}}{c_{PCM}} \; for \; T_{PCM} \geq T_{st,m} \end{cases} \quad (3)$$

where $T_{s,m}$ [°C] and $T_{st,m}$ [°C] are the melting temperatures at the start and stop points, respectively. In equation (3), H_{PCM} is expressed in kJ/kg. A similar system can be developed for the solidification process. The coefficients y_0, $A_{1,2}$ and $C_{1,2}$ are the results of a curve fitting process carried out in OriginLab. The values of the coefficients for solidification/melting are: y_0=24.84/30.82, A_1=-8.36/-19.40, A_2=-17.43/-15.14, C_1=-17.08/-19.91 and C_2=-70.05/-216.64.

4. The experimental setup

An experimental latent heat thermal storage was used to generate data for validation. The setup consists of a cylindrical evacuated double glass wall housing a row of 11 spheres containing RT21, an air-cooling system based on Peltier modules 136 W and cooled with air and water, a DC powered ventilator and a hot air gun. Figures 3 and 4 present the setup in schematic form and in the laboratory, respectively.

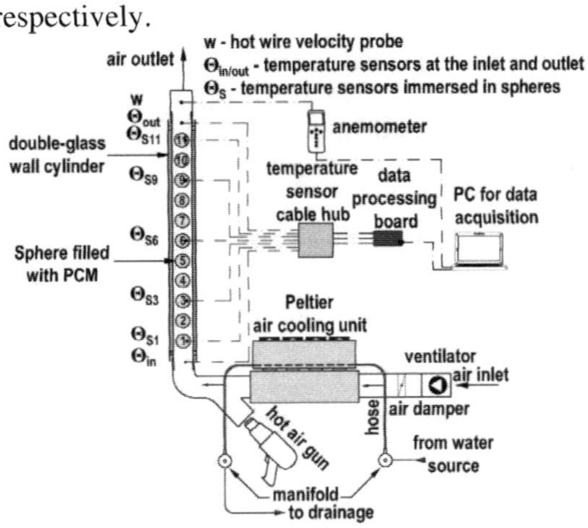

Figure 3. Schematic representation of the setup

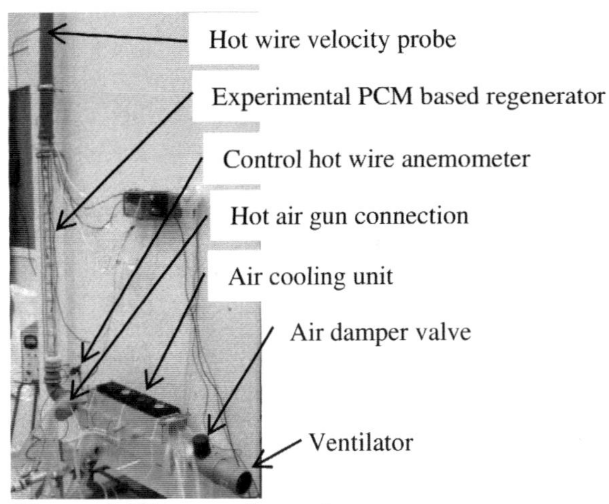

Figure 4. The experimental setup

A previous version of the setup was presented in [11].

IOP Conf. Series: Materials Science and Engineering **876** (2020) 012003 doi:10.1088/1757-899X/876/1/012003

5. Results and discussions

Two solidification experiments were carried out: at the air velocities of 0.65 and 0.80 m/s and the mean inlet air temperature after stabilization of 2.20 and 2.13 °C, respectively. Two melting experiments were carried out at the air velocities of 0.75 and 1.02 m/s and the mean inlet air temperatures after stabilization of 44.98 and 38.46 °C, respectively. The numerical values were compared with the measured data. The deviations were evaluated by the following metrics: mean bias error (MBE) and root mean square error (RMSE). Figures 5 and 6 present the graphical validation of the model for the highest and lowest values of the MBE and RMSE at solidification. The measured air inlet temperature (Θin) is also represented.

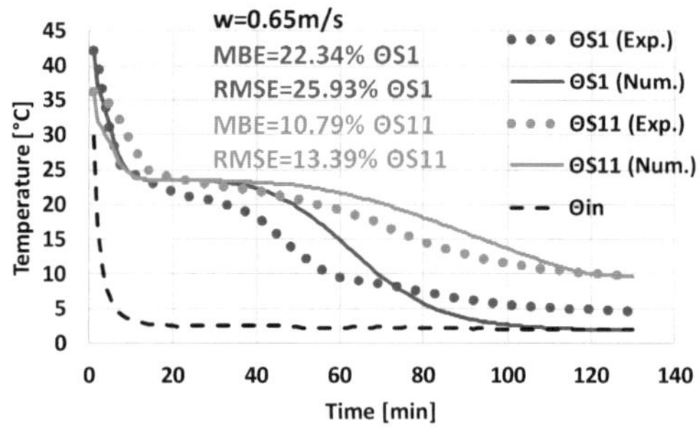

Figure 5. Measured and modelled values for solidification at an air velocity of 0.65 m/s

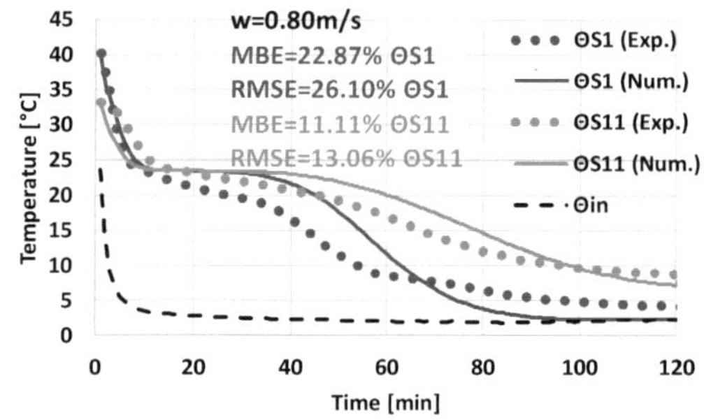

Figure 6. Measured and modelled values for solidification at an air velocity of 0.80 m/s

The numerical data is in good agreement with the measured values. The best agreement is achieved at the last sphere (ΘS11) as opposed to the first sphere where the highest deviations were observed. Better agreement is achieved at lower air velocities.

Figures 7 and 8 present the graphical validation of the model for the highest and lowest values of the MBE and RMSE at melting. The measured air inlet temperature (Θin) is also represented.

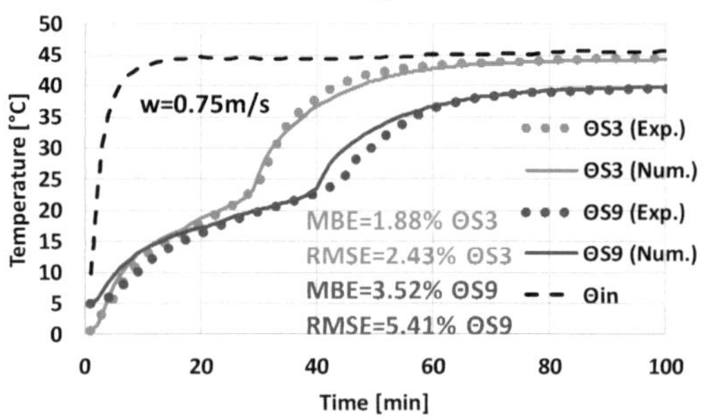

Figure 7. Measured and modelled values for melting at an air velocity of 0.75 m/s

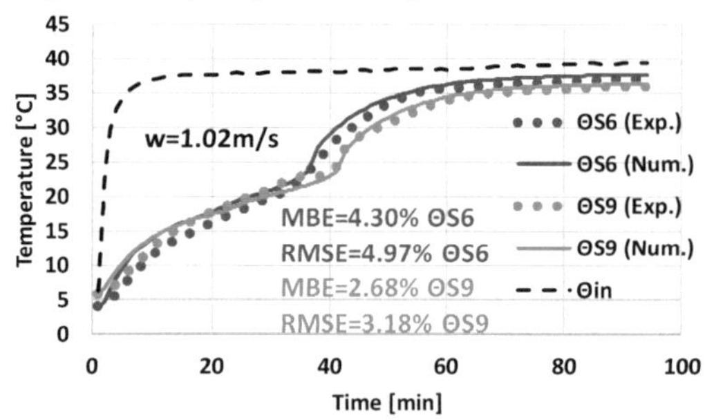

Figure 8. Measured and modelled values for melting at an air velocity of 1.02 m/s

The numerical data is in good agreement with the measured values.

Figure 9 presents the values of the RMSE and MBE for both solidification and melting.

It can be observed that the errors in the case of melting are significantly lower than in the case of solidification for all the considered air velocities.

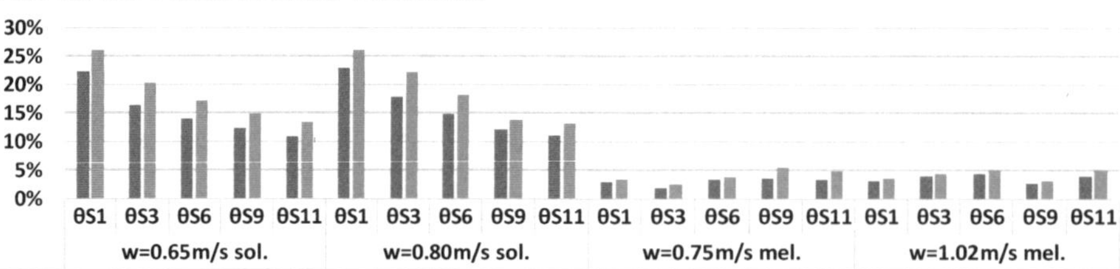

	w=0.65m/s sol.					w=0.80m/s sol.					w=0.75m/s mel.					w=1.02m/s mel.				
	ΘS1	ΘS3	ΘS6	ΘS9	ΘS11	ΘS1	ΘS3	ΘS6	ΘS9	ΘS11	ΘS1	ΘS3	ΘS6	ΘS9	ΘS11	ΘS1	ΘS3	ΘS6	ΘS9	ΘS11
MBE [%]	22.3	16.4	13.9	12.2	10.7	22.8	17.8	14.8	12.1	11.1	2.91	1.88	3.30	3.52	3.27	3.02	3.81	4.30	2.68	3.88
RMSE [%]	25.9	20.2	17.2	14.8	13.3	26.1	22.2	18.2	13.7	13.0	3.23	2.43	3.74	5.41	4.69	3.52	4.32	4.97	3.18	4.86

Figure 9. The MBE and RMSE values

In the case of solidification, it can be observed that the highest values of the deviations are for the first sphere, near the inlet, and the lowest values are for the last sphere, near the outlet. The errors decrease towards the outlet. No error variation trend can be identified in the case of melting.

6. Conclusions

A mathematical model was developed that employs the enthalpy method in order to estimate the temperature variations of PCM encapsulated in spherical shells. The thermophysical properties of the PCM were determined using the DSC dynamic method. Inverse enthalpy-temperature functions were obtained based on the measured data in order to take into account the thermal behaviour of the PCM during phase change. The model also takes into account natural convection during melting. The model was found to be in good agreement with the measured data. The best agreement was found at melting. Further accuracy analyses for different PCMs and at different air velocities and temperatures will be carried out in future studies. Future research will also be focused on using the mathematical model in order to study the thermal behaviour of latent heat regenerators in different operating conditions.

7. References

[1] Leite da Cunha S R and Barroso de Aguiar J L 2020 Phase change materials and energy efficiency of buildings: A review of knowledge Journal of Energy Storage 27:101083
[2] Klimeš L, Charvát P, Joybari M M , Zálešák M, Haghighat F, Panchabikesan K, El Mankibi M and Yuane Y 2020 Computer modelling and experimental investigation of phase change hysteresis of PCMs: The state-of-the-art review Applied Energy 263:114572
[3] Dutil Y, Rousse D R, Salah N B, Lassue S P and Zalewski L 2011 A review on phase-change materials: Mathematical modeling and simulations Renewable and Sustainable Energy Reviews, 15:112–130.
[4] Dutil Y, Rousse D, Lassue S, Zalewski L, Joulin A, Virgone J, Kuznik F, Johannes K, Dumas J-P, Bédécarrats J-P, Castell A and Cabeza L F 2014 Modeling phase change materials behavior in building applications: Comments on material characterization and model validation Renewable Energy 61:132-135
[5] Arkar C and Medved S 2005 Influence of accuracy of thermal property data of a phase change material on the result of a numerical model of a packed bed latent heat storage with spheres Thermochimica Acta 438:192–201
[6] Iten M, Liu S, Shukla A and Silva P D 2017 Investigating the impact of Cp-T values determined by DSC on the PCM-CFD model Applied Thermal Engineering 117:65–75
[7] Liao Z, Xu C, Ren Y, Gao F, Ju X and Du X 2018 A novel effective thermal conductivity correlation of the PCM melting in spherical PCM encapsulation for the packed bed TES system Applied Thermal Engineering 135:116–122
[8] Dermardiros V, Chen Y and Athienitis A K 2015 Modelling of an active PCM thermal energy storage for control applications Energy Procedia 78:1690 – 1695
[9] Egolf P W and Manz H 1994 Theory and modeling of phase change materials with and without mushy regions International Journal of Heat and Mass Transfer 37:2917–2924
[10] Cao E 2010 Heat transfer in process engineering ed T.M.-H. Companies
[11] Pop O G, Iuga C A, Fechete Tutunaru L and Balan M C 2018 Modeling and Experimental Validation of the Thermal Behavior of PCM using DSC Input Data American Institute of Physics Conference Proceedings 2004:020005

Acknowledgment

The research activities were supported by the project POCU/380/6/13/123927 – ANTREDOC, "Entrepreneurial competencies and excellence research in doctoral and postdoctoral study programs", project co-funded from the European Social Fund through the Human Capital Operational Program 2014-2020.

IOP Conf. Series: Materials Science and Engineering **876** (2020) 012004 doi:10.1088/1757-899X/876/1/012004

Finite Element Modelling of Big Turbo-Generator Rotor Vibrations using Newmark-beta Integration Method

Tarun Kumar[a*], Rajeev Kumar[a], S C Jain[a]

[a] School of Engineering, Indian Institute of Technology Mandi, Mandi, Himachal Pradesh, India

[*] kvtarun85@gmail.com

Abstract. In this paper big turbo-generator rotor vibrations are simulated for various electrical faults. Newmark-beta method is used to simulate the vibrations at various critical sections of the rotor. Rotor is modelled using finite element method with solid, hollow and tapered elements. Torque in the generator during the course of electrical faults is modelled using dq0 formulation and solved in MATLAB using 4th order predictor-corrector method. Line to ground (L-G), line to line (L-L), three phase (L-L-L) and asynchronization faults are simulated for 300msec. The torsional vibrations in rotor at critical section is found to be most severe during three phase fault followed by mal-synchronization fault followed by the line to line fault followed by line to ground fault.
Index Terms--Finite Element Modelling, dq0 model, Synchronous generator modelling, torsional vibration, turbo-generator rotor

1. Introduction

Torsional vibrations emerge in big turbo-generator rotor due to various electrical faults on transmission line can results in complete failure of the rotor and sometimes leads to the fatal failure. Therefore, it is essential to measure these torsional vibrations so that further effective control mechanisms can be developed. To measure torsional vibration on the rotor due to various electrical faults various researchers use analytical relation given in [1]. In [2-3] rotor is modelled using lumped mass model and analyzed for vibrations under unloaded state of the generator.

Different researchers use lumped mass model to analyze the big turbo-generator rotor and simulate the torsional vibrations under unloaded state of the generator using analytical relation for generator torque under various faults. Reference [4] uses dq0 formulation to model the torque during various electrical faults under unloaded condition of generator. Similarly, reference [5] use finite element model to analyze the big turbo-generator rotor with unloaded generator under faults. No comprehensive study is available in which torsional vibrations in the rotor during electrical faults has been analyzed under loaded conditions of generator. Therefore, a complete study of big turbo-generator rotor under line to ground, line to line, three phase and mal-synchronization faults is performed using finite element method. The rotor is modelled using solid, hollow and tapered cylindrical elements and coupled equations are solved numerically in MATLAB using Newmark-beta integration scheme.

2. Mathematical Formulation

2.1. Torque Modelling of Generator

The generator torque for line to ground and line to line torque is calculated using (1), (3) and (4).

whereas for three phase fault and mal-synchronization fault torque is calculated using (2), (3) and (4). Electromagnetic torque in generator during various faults is calculated using (5). The coupled turbine-generator equations are solved numerically using fourth order Adams predictor-corrector scheme with startup by fourth order Runge-Kutta method in MATLAB. The complete mathematical formulation for the torque during various electric faults is given in [4].

$$L.\frac{d}{dt}\begin{bmatrix} i_{phase} \\ i_{fd} \\ i_D \\ i_Q \end{bmatrix} + X\begin{bmatrix} i_{phase} \\ i_{fd} \\ i_D \\ i_Q \end{bmatrix} = \begin{bmatrix} u_{fd} \\ 0 \\ 0 \\ 0 \end{bmatrix} \tag{1}$$

$$L\cdot\frac{d}{dt}\begin{bmatrix} i_d \\ i_q \\ i_{fd} \\ i_D \\ i_Q \end{bmatrix} + X\begin{bmatrix} i_d \\ i_q \\ i_{fd} \\ i_D \\ i_Q \end{bmatrix} = \begin{bmatrix} u_d \\ u_q \\ u_{fd} \\ 0 \\ 0 \end{bmatrix} \tag{2}$$

$$\frac{d\omega}{dt} = -\frac{P}{J}T_{em} + T_{drive} \tag{3}$$

$$\frac{d\gamma}{dt} = -\omega \tag{4}$$

$$T_{em} = \frac{3}{2}\mathrm{p}\left(-L_{md}\, i_{md}\, i_q + L_{mq}\, i_d\, i_{mq}\right) \tag{5}$$

where, L and X are inductance and reactance matrices derived similarly as in [4], [6].

2.2. Finite Element Modelling of Rotor

Big turbo-generator rotor is modelled using finite element formulation. Different sections of rotor are analyzed using solid, hollow and tapered cylindrical elements. Hamilton's principle is used to formulate the finite element equations. The elemental equation for a particular element is given by (6). The elemental equations are combined to form global finite element equations for rotor dynamics as shown in (7).

$$[m]^{(e)}\{\ddot{\theta}\}^{(e)} + [k]^{(e)}\{\theta\}^{(e)} = [T_p] + [T_{unif}] \tag{6}$$

where,

$$[k]^{(e)} = \int_0^l GJ\begin{bmatrix} \left(\frac{dN_1}{dx}\right)^2 & \frac{dN_1}{dx}\frac{dN_2}{dx} \\ \frac{dN_1}{dx}\frac{dN_2}{dx} & \left(\frac{dN_2}{dx}\right)^2 \end{bmatrix}dx, \quad [m]^{(e)} = \int_0^l \rho J\begin{bmatrix} N_1^2 & N_1N_2 \\ N_1N_2 & N_2^2 \end{bmatrix}dx$$

ρ is density, J is polar moment of inertia, l is length, G is shear modulus, T_p is point torque and T_{unif} is uniform torque on the element. Appropriate damping matrix can be introduced to get final global finite element formulation.

$$[M]_{(g)}\{\ddot{\theta}\}_{(g)} + [C]_{(g)}\{\dot{\theta}\}_{(g)} + [K]_{(g)}\{\theta\}_{(g)} = [T_p]_{(g)} + [T_{unif}]_{(g)} \tag{7}$$

2.3. Dynamic Modelling of Coupled Turbo-generator Rotor

The coupled turbo-generator rotor dynamics is solved using Newmark-beta method. The electromagnetic torque calculated in (5) is applied as uniform torque on generator elements after multiplying it with the base. The rotor at the start of the fault is assumed to be rotating at 3000 RPM. The initial torsional displacement of rotor is assumed to be zero at the time of start of electrical fault. The coupled system is numerically solved in MATLAB.

3. Validation

The dynamic model for synchronous generator as described in section 2.1 is validated by the results from [4]. The values of inductance L_e and resistance R_e are assumed to be zero for validation purpose only. Since dynamic model is same for line to ground and line to line short circuit fault, the results are validated for generator torque in line to ground short circuit for the most unfavourable conditions given in [4] as shown in Fig.1(a).

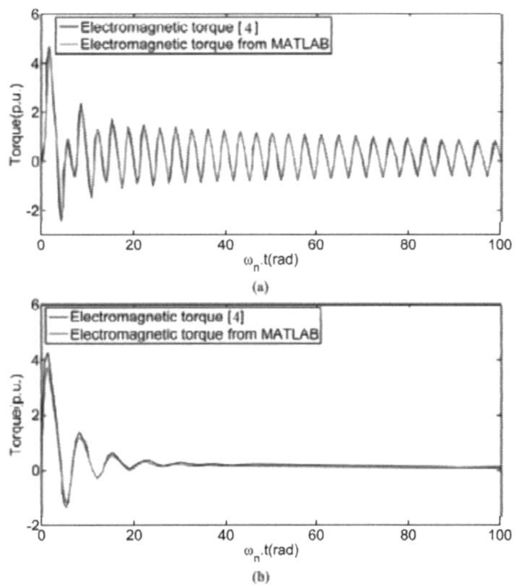

Figure 1. (a) Comparison of calculated electromagnetic torque during line to ground short circuit with [4] results. (b) Comparison of calculated electromagnetic torque during three phase short circuit with [4] results.

Table 1. Comparison of first three natural frequencies from MATLAB and Reference [7].

No of Elements considered in MATLAB	First 3 Natural Freqs. From MATLAB ($x10^3$Hz)	First 3 Natural Freqs. of Ref. [7] ($x10^3$Hz)	% Error	Avg. % Error
	1.682	1.534	9.647	
5	7.737	4.911	57.544	174.46
	46.247	8.315	456.187	
	1.576	1.534	2.737	
6	5.553	4.911	13.072	14.294
	10.566	8.315	27.071	
	1.557	1.534	1.499	
7	5.045	4.911	2.728	6.047
	9.472	8.315	13.91	
	1.551	1.534	1.108	
8	4.859	4.911	-1.058	1.904
	8.786	8.315	5.664	
	1.547	1.534	0.847	
9	4.774	4.911	-2.789	-0.302
	8.401	8.315	1.034	
	1.547	1.534	0.847	
10	4.773	4.911	-2.810	0.0353
	8.487	8.315	2.068	

Similarly, dynamic model is same for three phase short circuit and phase synchronization fault. So, results are validated for generator torque in three phase short circuit for most unfavourable condition as given in [4] and shown in Fig. 1(b). From Fig. 1(a-b), it is clear that numerical results are matching well with the [4] results.

Further the finite element model for the rotor is validated using [7] results. The first three natural frequencies are calculated to validate the global mass and stiffness matrices. The first three natural frequencies are compared with the [7] and convergence of the code is established in Table 1. The first three natural frequencies of the tube as given in [7] show close resemblance to the simulated results.

4. Results and Discussion

A coupled turbo-generator rotor with 187 elements are chosen for the vibration analysis. The synchronous generator rated 588 MVA power is connected to rotors of Low pressure, intermediate pressure and high-pressure turbine through couplings. Three types of elements viz solid, hollow and tapered elements are used to analyse the rotor.

The generator is modelled using dq0 formulation. The complete system is connected with the infinite load through grid with resistance R_e and inductance L_e. The generator torque vibrations are plotted in Fig. 2 for Line to ground, Line to Line, Three Phase and mal-synchronization faults. Per unit system is used to calculate the peak values of torque. The peak values of pu torque for line to ground, line to line, three phase and mal-synchronization faults are 12 pu, 7.9 pu, 17.1pu and 8.1pu respectively. The actual value of electromagnetic torque is calculated by multiplying per unit value by torque base $3pU_nI_n\omega_n^{-1}$. The various parameters of generator are given in Table 2.

Each element of rotor has two degrees of freedom i.e. one rotational degree of freedom at each node. Since torsional vibrations are independent from translational and axial vibrations so other DOFs are neglected to keep the formulation simple.

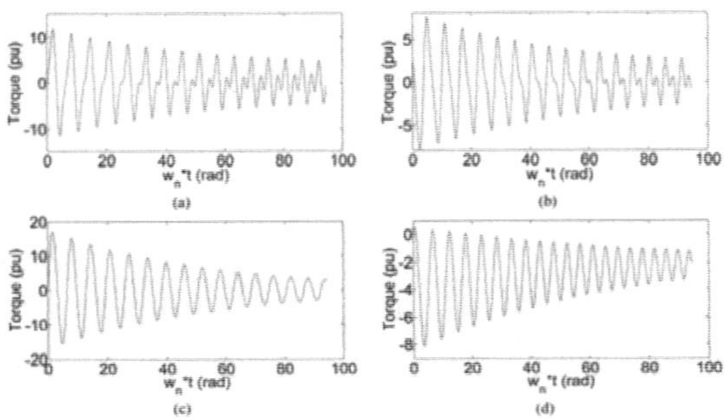

Figure 2. (a) Per unit generator torque during line to ground fault. (b) Per unit generator torque during line to line fault. (c) Per unit generator torque during three phase fault. (d) Per unit generator torque during mal-synchronization fault.

Table 2. Various parameters used in generator modelling.

Parameter	Symbol [a]	Value
Parameters related to generator	S_n	588 MVA
	$U_{line,n}$	11000 V
	$Cos\varphi_n$	0.85
	n_n	3000 rpm
	f	50 Hz
Stator winding parameters	R	0.00193 p.u.
	L_σ	0.1256 p.u.
Rotor windings parameters	R_{fd}	0.0009 p.u.
	R_D	0.0093 p.u.
	R_Q	0.008 p.u.
	$L_{fd\sigma}$	0.02721p.u.
	$L_{D\sigma}$	0.33 p.u.
	$L_{Q\sigma}$	0.08 p.u.
d-q axis magnetizing inductances	L_{md}	1.9539 p.u.
	L_{mq}	1.8504 p.u.
Line resistance	R_e	0.1235 p.u.
Line inductance	L_e	6.773 p.u.
Infinite Voltage	U_∞	8.7639 p.u.

a -Various symbols and related subscripts are given in [4]

Torsional vibration in the rotor are simulated for element 39. The electromagnetic torque variation during any electrical fault is applied on element 132-140. The initial speed of rotor is assumed to be 3000 RPM and initial torsional displacement is assumed to be zero. The vibrations in element 39 during any fault is plotted for 100msec. The finite element equation is solved using Newmark-beta integration method.

During the fault time the different sections of the rotor rotate with different angular velocity. Due to this, different sections of the rotor receive different peak value of stresses. The coupling joint between various turbines and generator are most probable sections to be failed during the faults. So, coupling element 39 is analysed numerically for vibrations produces during various electrical faults.

To establish the stability of the system, the angular displacement Vs angular velocity is plotted for element 39 as shown in Fig. 3. It is clear from Fig. 3(a-d) that nonlinear system of equation is stable during the simulation time.

The torque in element 39 during line to ground, line to line, three phase and mal-synchronization fault are plotted in Fig. 4 (a-d). The peak torque during the line to ground, line to line, three phase and mal-synchronization fault is 0.72MNm, 0.54MNm, 1.1MNm and 0.55MNm respectively. It is clear from Fig. 4 that most severe fault is three phase faults, followed by line to ground fault, followed by mal-synchronization fault, followed by line to line fault. The frequency of torque is maximum in case of mal-synchronization fault as compare to the three-phase fault. So, during mal-synchronization fault, rotor can lead to fatigue failure earlier.

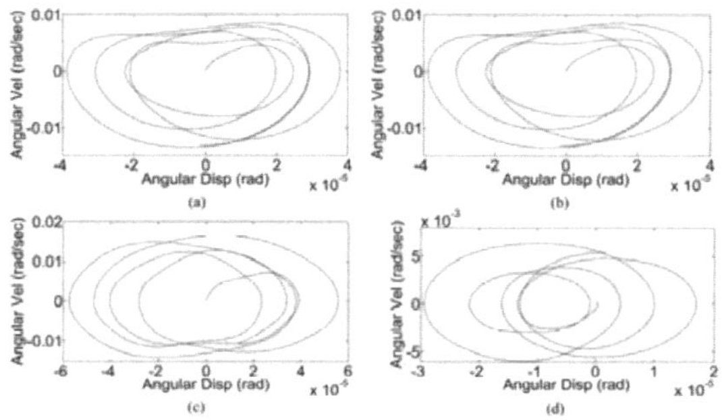

Figure 3. Relative angular displacement Vs relative angular velocity plot for element No. 39. (a) Line to ground fault (b) Line to line fault (c) three phase fault (d) Mal-synchronization fault.

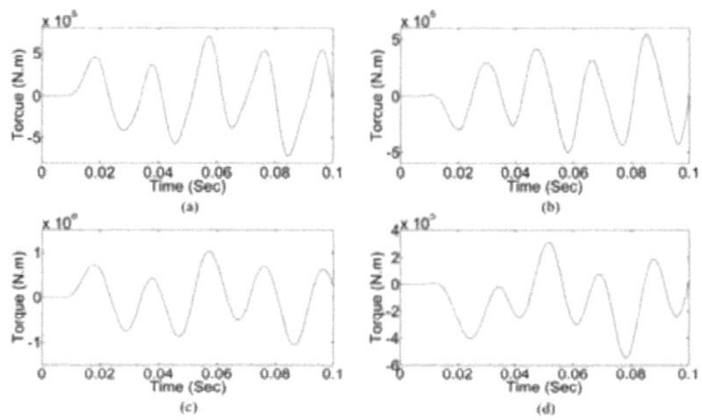

Figure 4. Torque in element No 39 during various electrical faults. (a) Line to ground fault (b) Line to line fault (c) three phase fault (d) Mal-synchronization fault.

5. Conclusions

The dynamic models of loaded synchronous generator under four types of electric faults are analysed in MATLAB. The rotor is modelled using FEM and complete dynamic system is numerically simulated using Newmark-beta integration method in MATLAB. The most severe torque observed during three phase fault followed by the line to ground, mal-synchronization and line to line fault. Further, numerical study can be performed to analyse the various critical section of the rotor for the peak torque and stress. An active control mechanism can be developed by analysing the numerical results.

6. References

[1] Wachel JC, Szenasi FR. "Analysis of torsional vibrations in rotating machinery". In Proceedings of the *22nd Turbomachinery Symposium 1993. Texas A&M University*. Turbomachinery Laboratories.

[2] Cudworth, C. J., and J. R. Smith. "Steam turbine generator shaft torque transients: a comparison of simulated and test results." In *IEEE Proceedings C (Generation, Transmission and Distribution)*, vol. 137, no. 5, pp. 327-334. IET Digital Library, 1990.

[3] Chyn, C., R-C. Wu, and T-P. Tsao. "Torsional fatigue of turbine-generator shafts owing to network faults." *IEE Proceedings-Generation, Transmission and Distribution* 143, no. 5 (1996): 479-486.

[4] Lupsa-Tataru, Lucian. "Comparative simulation study on synchronous generators sudden short circuits." *Modelling and Simulation in Engineering* 2009 (2009): 8.

[5] R. Oliquino, S. Islam, and H. Eren, "Effects of Types of Faults on Generator Vibration Signatures," in *Australasian Universities Power Engineering Conference*, 2003, pp. 1–6.

[6] Kumar, T., Bangunde, A., Kumar, R., & Jain, S. C. (2018, June). The Torque Simulation of Synchronous Generator Under Loaded Condition for Different Types of Electrical Disturbances. In *2018 2nd International Conference on Power, Energy and Environment: Towards Smart Technology (ICEPE)* (pp. 1-9). IEEE.

[7] Sung, Chia-Chi, Vasundara V. Varadan, Xiao-Qi Bao, and Vijay K. Varadan. "Active torsional vibration control experiments using shear-type piezoceramic sensors and actuators." *Journal of intelligent material systems and structures* 5, no. 3 (1994): 436-442.

Analysis of different microcracks shapes and the effect of each shape on performance of PV modules

Mathhar Bdour[1*], Aseel Al-Sadi[2]

[1] Department of Energy Engineering, School of Natural Resources Engineering and Management, German Jordanian University, PO Box 35247, Amman 11180, Jordan.
[2] Philadelphia solar, Al Qastal Industrial Area 2. Airfreight Road. Amman, Jordan. P.O. Box 143808, 11814 Amman, Jordan.

*Corresponding author: Mathhar Bdour: madher.bdour@gju.edu.jo

Abstract. Photovoltaic solar panels became the world's largest distributed Renewable technology through its easy manufacturing and installation, moreover, the cost of photovoltaic panels falls continuously which provided a safe environment to investors to invest in such systems. The Photovoltaic solar panel made from different layers starting from anti reflected coated glass layer, encapsulation material, solar cells matrix another encapsulant layer and back sheet. This structure does not provide the needed protection to the solar cells encapsulated inside and there will be a possibility of microcracks happen at different stages of the project life cycle starting from the production, transportation, installation, cleaning and operation, and maintenance stage. Through this research, modules from different projects were tested to see the microcracks and their effect. Results show that those cracks can cause a power drop up to 3.21% depending on the type of the module. Furthermore, microcracks are classified as minor and major ones based on their shape, dendritic and vertical shapes are resulted to be the most severe ones.
Keywords. Microcracks, solar energy, power drop.

1. Introduction

The need for analysis of microcracks is important since numerous power plants were already installed and started to show some problems, mostly power drop due to microcracks. Taking into consideration that changing any panel from a project without performing a root cause analysis diagram and finding the effect on the long term is a real loss firstly to the project owner and secondly to the environment which is our top priority. Jordan is the second most attractive environment for renewable energy and energy efficiency investments in the region with only 1.3 millions of tons of Carbon Dioxide per terawatt hours
(MtCO2/TWh) in CO_2 emissions from fuel combustion/electricity output, the country is doing relatively well in the energy sector despite a lack of natural resources additionally 100% of Jordan's total population has access to electricity and access to clean fuels and technology for cooking. The energy market is now poised to become a leading regional example of non-oil dominated energy mix and diversified energy sources [1]. According to the Ministry of Energy and Mineral Resources Report (2018) which referred to (1130 MW) installed capacity constituted 11 percent from total electricity generation capacity in Jordan and this capacity is expected to increase to 2,400 megawatts (MW) by 2021 as highlighted in the 2018 report [2]. If that plan is applied carefully and the expectations are

achieved, then the renewable energy would constitute 20 percent of Jordan's total electricity generation capacity compared to 1 percent in 2014.
According to the ministry's report, renewable energy projects that are set for completion in 2021 will have a total capacity of 1,270MW. Photovoltaics (PV) have been adopted as a major renewable energy source globally, for instance, India plans to reach up to 60 GW by 2022 [3]. Furthermore, entering the industry of electric vehicles [4]. The construction of PV systems is capital intensive and systems are expected to perform and generate power for more than 25 years [5]. PV solar modules consist mainly from solar cells and other components encapsulated to each other's and many types of PV solar cells can be used to assemble PV solar modules, Mono-Crystalline (MONO-SI), Poly or Multi-Crystalline (P-Si), Amorphous Silicon (A-Si), Concentrated PV cell (CVP).
The thickness of solar cells (Poly or Mono) varies from supplier to another and the record lab cell efficiency is 26.7% for mono-crystalline and 22.3% for multi-crystalline silicon wafer-based technology. The highest lab efficiency in thin-film technology is 23.4% for CIGS and 21.0% for CdTe solar cells [6].
Material usage for silicon cells has been reduced significantly during the last 13 years from around 16 g/Wp to about 4 g/Wp due to the increased efficiencies, thinner wafers and wires as well as larger ingots [6]. Discussing the overall PV solar panels structure and after assembling all layers as shown in Figure 1 some types of microcracks may grow according to their shape and location in PV module and this grow might happen due to, transport, improper installation, handling, vibration, excessive load, environmental stresses, improper cleaning and during the operation and maintenance stage. It is important to find the impact of each microcrack shape on the performance of the PV module to optimize the power generation from the PV project and reduce the cost of operation and maintenance thereafter.

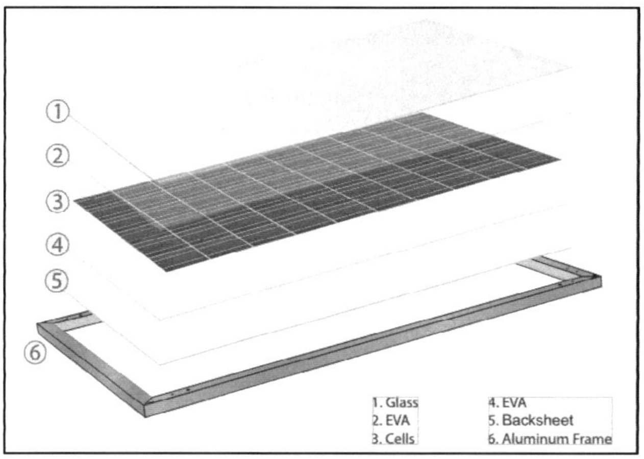

Figure 1. PV solar module structural layers.

The orientation of the microcracks plays an important role in assessing their potential for causing power loss. Microcracks parallel to the cell busbars (vertical) are critical, also dendritic cracks and cracks with multiple orientations. Considering that on average 6% of modules can show cracks after transportation, it is important to assess the type and severity of microcracks before acceptance and installation of modules into a PV system. The shape parameters of the micro-crack pattern are dependent on the EL measurement conditions, the most important of which are the bias current and camera exposure level [7]. Cracks or micro-cracks played a key role in explaining this mechanism, they prevented the moisture from evaporating out of the module, and narrow cracks/ micro-cracks are sturdier than soldered cut lines in preventing moisture getting out [8]. It also explains why there were no snail trails at the edges of cells; they may appear if the humidity is high enough inside EVA and Ag grid lines [8]. This Study will focus on certain types of microcracks which are vertical, horizontal, dendritic, and others such as (PID dark cells on edges, dark cells, and Impact μCracks).

IOP Conf. Series: Materials Science and Engineering **876** (2020) 012005 doi:10.1088/1757-899X/876/1/012005

2. Testing and data analyzing

In this work, an analysis of microcracks was performed to find the effect of each microcrack shape on the performance of the PV module. The affected modules disassembled from the project site and returned to the workshop to measure the EL image and analyze the microcrack then I -V curve measured by flasher machine to find the drop in power performance. The EL tester used is equipped with NIR Nikon Camera with HD lens and total resolution ≥16.2 Megapixel and image pixel 405μm /pixel with testing Cycle of 30 s/Module and this machine is suitable for both monocrystalline silicon and polycrystalline silicon module.

The test procedure was done by loading the module into a loading conveyor then positioned in a way that the camera detects all cells while the EL detector receives the upstream application signal. If the EL detector, probes contact electrodes automatically then the test is done. After testing done the module ejected to offloading conveyor and the images saved to be analyzed. While the Sun Simulator machine was used to test the power and electrical specification, the machine is equipped with Endeas software a class AAA complied with IEC60904-09 concerning spectrum, irradiance non-uniformity, and short-term instability (STI). The Xenon technology spectrum is continuous from 300 nm up to 1,200 nm.

The measurements were done following IEC 61215-2 [9], where module shall be loaded into Sun simulator (flasher) class AAA and a PV reference device by IEC60904-2 shall be used as a reference module of the same size with the same cell technology to match spectral responsivity. Then the Current-voltage characteristics of the sample/module determined following IEC60904 -1 at a specific set of irradiance and temperature conditions.

3. Data Analysis

The manufacturing processes might affect the cells and cause different types of microcracks. The most effective shape is dendritic when the cell comes darker in color, it also noticed that the more dendritic darks cells the more power drop recorded. The key role for power drop is having dark areas in the microcrack itself which makes that area non-active and in dendritic shape, the microcracks spread along with the cells and cut more than one bus bar. In this study, the microcracks during the manufacturing process are neglected and the concentration will be on them after installation. Different projects were analyzed in different locations to find the effect of microcracks and summed up like Figure 2. It was noticed significant power drop in the first module due to having many types of microcracks at the same time besides having dark cells which reduced the power significantly as the cells connected in series in the PV module. Whereas the second module showed significant power drop as well due to having many types of microcracks at the same time and on top of that it had dark areas in two cells. The LID effect for Mono-Crystalline Modules is up to 3 % which is doubled the poly-Crystalline modules. EL images and flasher results of a total of 44 crystalline silicon modules in different locations were taken during the 2019 study. The images and electrical data refer to solar panels not yet installed, others installed in Dabouq, ALQastal, Cairo, AlMafraq, Ein ELBasha, and one transported from China to Amman.

1. Dendritic with dark areas Disconnection for current (Major).

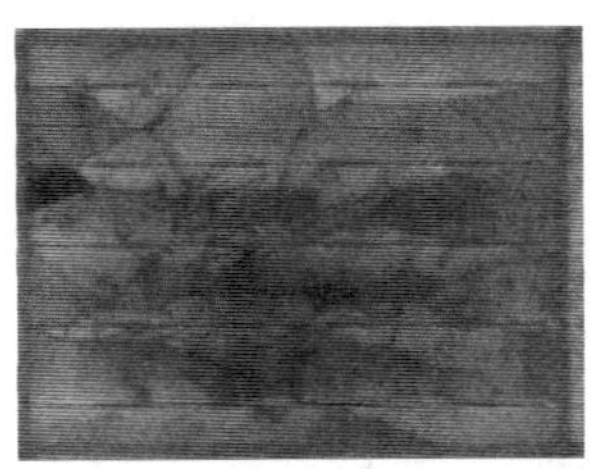

2. Dendritic with no dark area but still (Major) cut all busbars.

IOP Conf. Series: Materials Science and Engineering **876** (2020) 012005 doi:10.1088/1757-899X/876/1/012005

3. Vertical with dark area, disconnection of current (Major).

4. Vertical on busbar, per number per each cell (Major).

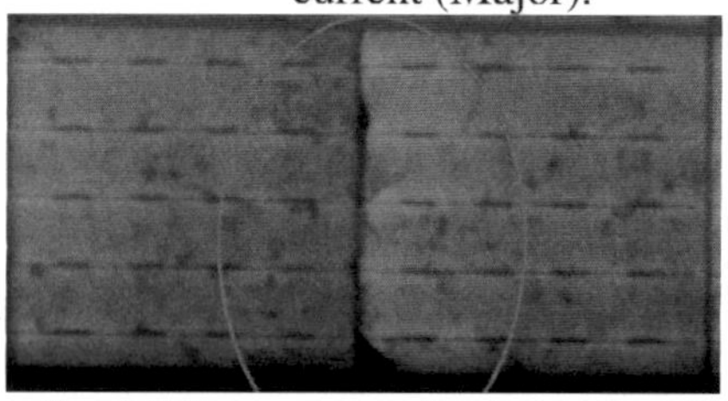

5. V shape on one busbar out of five busbars, cause degradation on long run (Minor).

6. Vertical cutting all busbars (Major).

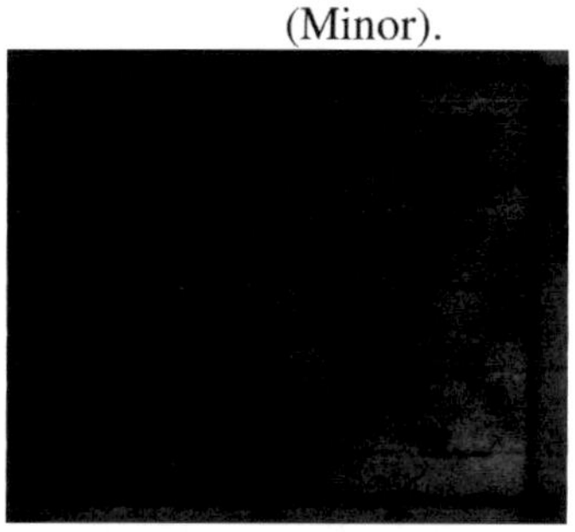

7. PID effect on 90% of the cell (Major)

8. Mechanical damage on 20% (Major).

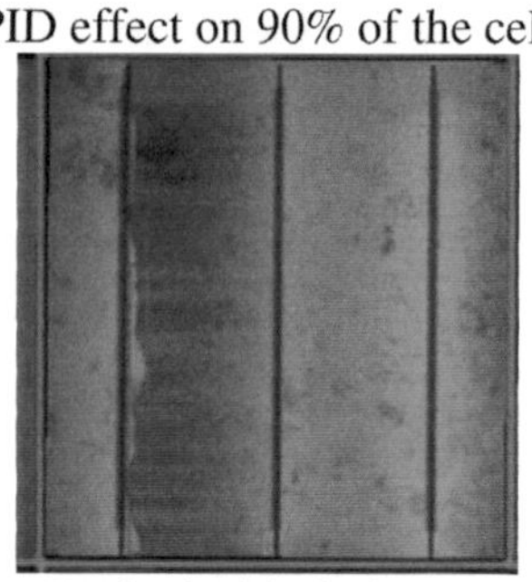

9. Vertical with dark area (Major).

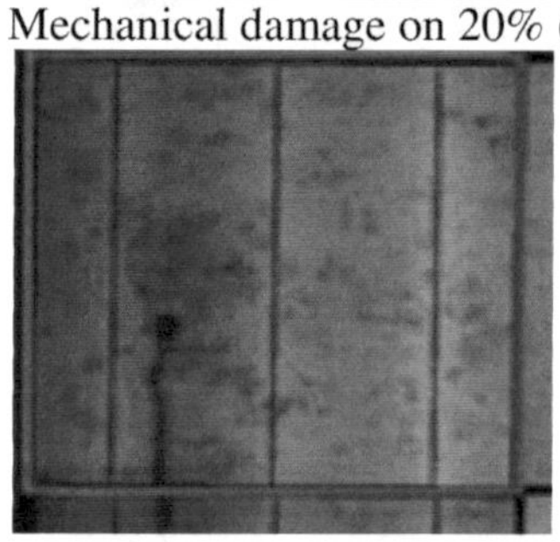

10. Back sheet scratch (Major).

11. Impact and it can be expanded (Major).

12. Horizontal crack (not critical) (Major).

Figure 2. Micro cracks classifications from several projects

Microcracks testing was performed on projects installed in 2016 and 2019, also for poly and monocrystalline cells. Generally, microcracks affect the power generated with an average drop of 1.49% in poly-crystalline modules and 0.55% in mono-crystalline ones. It is noticed that older projects cracks effects are higher than new ones, this is expected because old projects are more vulnerable to weather conditions and other stresses.

4. Conclusion

The quality of PV power plants and projects is very critical to generate the needed power production, long life of the project and the return of investment which investors look for, therefore analyzing the micro cracks is needful to get the optimum benefit. During the installation, operation, and maintenance of the power plant, some kind of defects might happen and in this case, each type of defect must be defined to find the seriousness on the power plant as some of them might cause a significant drop on power and create safety issues afterward. There are many ways of developing microcracks in PV modules starting from the manufacturing process, transportation, improper installation, and improper cleaning.

It is also important to point out that the snail trail appeared visually in some samples mentioned in this study grown in the path of microcracks where every microcrack shown in such samples had snail trail but not grown nor increased by time as long as the source of microcracks eliminated.

Therefore snail trail is visual appearance correlated to microcracks but it will not appear if there is no source of micro cracks and there were many research and development done on this matter to overcome this problem and hide the snail trail but still the microcracks have existed as there are many reasons of developing them. Microcracks effect can be classified as a major one on the module if it is for example in vertical shape, power reduction can increase significantly. Therefore, this study aims to classify the micro cracks to know whether there is a need to change the modules or not. Power reduction is expected accordingly where values vary 0.84% to 3.21% power drop in poly-crystalline modules, and 0.22% to 0.89% in mono-crystalline ones.

5. References

[1] PI Berlin 2019 *Industry Trends in PV Module Quality from over 250 Factory Audits*

[2] Anon 2018 *Annual report, Ministry of energy and mineral resources-Jordan*

[3] Department of Electrical Engineering, National Institute of Technology Rourkela, Dist. Sundargarh, Odisha-769008, India, Tarai R K and Kale P 2016 Validation of predictive models to estimate annual PV production: A case study of odisha *ijsgce*

[4] Department of Electrical and Electronic Engineering, Kyushu Institute of Technology, Kitakyushu 804-8550, Japan, Kawagoe T, Fuji K, Shiota A, Mitani Y and Qudaih Y S 2014 Construction of power supply system using electric vehicle for stable power delivery from PV generation *ijsgce*

[5] Anon 2019 *An Effective Method On Evaluating Photovoltaic Module Snail Trail* (in SNEC, China)

[6] Dubey R, Vasi J, Chattopadhyay S, Zachariah S, Rambabu S, Singh H K, Kottantharayil A, Arora B M, Narasimhan K L and Shiradkar N 2018 On-Site Electroluminescence Study of Field-Aged PV Modules *2018 IEEE 7th World Conference on Photovoltaic Energy Conversion (WCPEC)* 2018 IEEE 7th World Conference on Photovoltaic Energy Conversion (WCPEC) (Waikoloa Village, HI: IEEE) pp 0098–102

[7] Dhoke A, Sharma R and Saha T K 2018 PV module degradation analysis and impact on settings of overcurrent protection devices *Solar Energy* **160** 360–7

[8] Commission J I 2018 *Energy sector profile*

[9] F. ISE 2019 *Photovoltaics Report* (Fraunhofer ISE, Frieburg)

Acknowledgments

The research conducted within this study has been supported by the German Jordanian University in Amman, Jordan.

Study of printed polymeric flexible energy harvesting elements – impact of the electrode materials and patterns

M Aleksandrova[1], R Aepuru[2], G Dobrikov[1]

[1]Technical university of Sofia, Department of Microelectronics, Sofia, Bulgaria
[2]University of Concepcion, Department of Materials Engineering, Concepcion, Chile

m_aleksandrova@tu-sofia.bg

Abstract. In this paper is presented study of flexible pressure piezoelectric energy harvesting developed with an screen-printing of polyvinylidene fluoride-trifluoroethylene (P(VDF-TrFE)) ink on three different metal electrodes (silver, gold and aluminium) with three different topologies – rectangular, meander and side comb. The influence of the electrode type and pattern is studied in terms of polymer coating distribution and morphology, piezoelectric voltage generation and stability at multiple bending, as well as piezoelectric coefficient. The elements are studied at low mass loading of up to 100 g and frequency up to 50 Hz. They produce voltage between 916 and 362 mV from active area of 1 cm^2 and piezoelectric polymer film of 3 μm, according to the electrode metal nature and shape. It was found that the silver electrode with meander shape is the optimal for electrical performance and mechanical stability giving superior piezoelectric coefficient of 0.332 V.m/N to the reported values and only 8.6 % decrease of the piezoelectric voltage at 1500 repeating bending cycles.

1. Introduction

Nowadays, the demands on the portable, thin and light-weight energy harvesting devices from the ambient processes continuously increase, because they tent to serve as alternatives of the conventional batteries in the miniaturized devices (nanoelectronic or nanoelectromechanic). In contrast to the macro-piezoelectric harvesters, where loading of tenth of kilograms need to be stand and few Watts power are generated as a result, the typical efficiency of the single piezoelectric harvesting unit with area of few square centimetres and films thickness about 1-2 μm is low (not greater than few hundreds of nW or maximum few μW at higher thickness and depending on the dielectric permittivity of the piezoelectric materials) [1,2]. Therefore, few general approaches are usually applied in order to gain the piezoelectric response, such as: multilayer stack structure with intermediate electrodes, coupled in parallel; composites between piezoelectric oxides, each with high piezoelectric modulus and dielectric permittivity, multiple coated substrates connected in parallel (or series), and etc. [3]. Due to the requirements for environmental friendly technologies, lead-free, strontium-free and zirconium-free materials are recently preferred for energy device fabrication. They are characterized with much lower piezoelectric coefficients compared with the rest piezo-oxides. Piezoelectric polymers based on PVDF (polyvinylidene fluoride) replace the lead-free oxides because of the similarity in the piezoelectric modulus and dielectric permittivity, but advantageously possesses high strain resistance and great durability at bending/twisting. Additional consideration should be the high sensitivity of the piezoelectric transducers, because some of the typical applications related to health care uses mechanical stimuli with equivalent mass load not greater than 100-200 g and low frequency (for example hearth beating and breathing processes).

Recently, a lot of reports can be found about ferroelectric composite oxide-polymer or purely polymer-based transducers [4-6]. Some of the researches are devoted to chemical synthesis of new materials of chemical modification of the existing ones for enhancement of the piezoelectric constant and durability [7,8]. Another reports aim to tune the electrical properties, such as increase of the permittivity, decrease of the loss factor, increase the dipole mobility, polarization, and etc. [9,10]. In the same time, information lacks about the influence of the metal electrodes on the electrical charge collection ability and compatibility with the piezoelectric materials processing technologies (deposition and annealing).

The poly[(vinylidenefluoride-co-trifluoroethylene] (PVDF-TrFe) is modification of the conventional PVDF ensuring better ferroelectric properties and enhanced piezoelectric constant than the PVDF [11], therefore, the materials has attracted the researchers' attention almost ten years ago, when it has been incorporated as a coating in the non-volatile memory cells [12]. The next promising applications reported more recently are inkjet-printed piezoelectric actuator with polymer P(VDF-TrFE) thin film [13] and flexible piezoelectric transducers relying on thin 200 nm polymer functional film [14]. Based on the numerous studies, one can be concluded that the material has favourable polarizing properties, good processing ability and good compatibility with the most of the used materials and technologies in the microfabrication.

In some typical micro- and nanoelectromechanical elements like sensors, or actuators, the involved metals and geometrical patterning of the electrodes are crucial for the optimization of the device electrical performance. Therefore, study of the impact of the electrodes material and configuration is strongly necessary. The literature overview showed that there is no complete investigation related to this issue. For the application of piezoelectric polymer PVDF-TrFe reports, showing the effect of the electrodes on the piezoelectric performance has not been found. Moreover, for flexible energy harvesting based on PVDF-TrFe additional considerations due to adhesion problems should be also taken into account, in addition to the technological compatibility with the substrate and piezoelectric materials.

In this paper, new formulated PVDF-TrFe ink was printed between three different electrode materials (gold, silver and aluminium) with three different patterns (square, meander and side-comb shapes (the last with non-mutually overlapped "fingers"). Comparison between the two different sets of samples was made in terms of piezoelectric output voltage and long-term stability of the signal. In this way, complex and expensive multilayer structures with inner electrodes are avoided and in the same time the collecting of the output charge is improved due to lack of internal coupling and interfacial strain problems.

2. Experimental section

Polyethylene naphthalate (PEN) plastic substrates 125 μm were cleaned and pieces were cut for bottom electrode deposition. The sizes were considered according to the calculated equivalent area of the films due to the different patterns. Three different metals were considered with small thickness in order to avoid ductility problems. Gold films with thickness of 50 nm and sheet resistance 4.5 Ω/sq was DC sputtered in vacuum at plasma current of 35 mA. Silver and aluminium films with the same thickness and sheet resistance of respectively 3.4 Ω/sq and 5.6 Ω/sq were thermally evaporated in vacuum. Standard photolithographic processes were applied - direct photolithography for the gold films with potassium iodide etching solution for patterning and reverse photolithography (lift-off process) in order to avoid etching in nitric acid and phosphorous acid, respectively for the silver and aluminium films. Square, meander and side-comb shapes were obtained. Suitable contact pads were produced in order to connect neighbouring segments in series and make the active functional area equal.

P(VDF-TrFE) ink was screen printed (printer model S300V) on metal coated PEN substrates, which were selected to avoid thermal stress during annealing of the piezoelectric coating. Screen with mesh count of 180 n/cm and wire diameter of 30 μm was used and the deposition parameters were squeegee speed of 200 mm/s and pressure of 3 bars. The annealing of the samples was done in furnace at 135°C per 15 minutes in order to stabilize the prints and guarantee interruption of the rheological behaviour of the print. At these deposition and annealing conditions it was found that homogeneously distributed and crystallized P(VDF-TrFE) films could be produced on any of the metals coating the plastic substrate without cracks and other defects. The film thickness was approximately 3 μm, measured by profilometer. After deposition and patterning of the top electrodes polling was performed by applying a constant electric field of 50 V/μm. Top electrodes of the same type with the same patterns were realized. Photos of the bottom electrodes and the completed prepared samples are shown in figure 1 a-c.

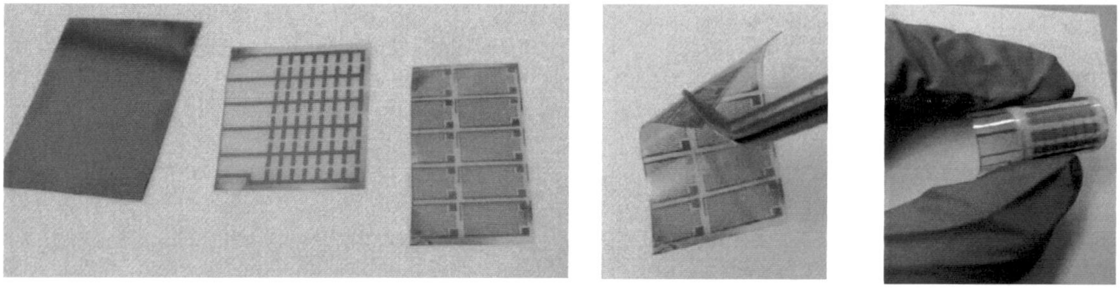

a) bottom electrode patterns b) flexed PEN/Au c) completed structure

Figure 1. Photos of the prepared samples.

The electromechanical testing setup (described elsewhere [15]) provides single point clips and free standing opposite side of the sample (cantilever type testing in dynamic mode), thus apply precisely controlled strain. The way of fixing and strain applying with respect to the electrodes places (capacitor type structure) results in a g_{31} measurement mode. The magnitude of the vibration equivalent to mass load provides maximum load of 100 g and broad range of frequencies from 2Hz to 2 MHz. The produced piezoelectric voltage was monitored by oscilloscope Tektronix TDS 1012B and the surface morphology of the coatings was observed by Dino-Lite digital microscope at magnification of 500. In order to gain the current produced from a unit element it was not tested single structure (except for the rectangular one), but multiple smaller elements were connected in parallel by the conductive lines and pads, shaped during the photolithography.

3. Results and discussion

The effect of the different metal electrodes and their shapes on the piezoelectric behaviour was compared and summarized in figure 2. It can be observed that no matter of the electrode pattern, the samples with silver electrodes produced the greatest piezoelectric voltage, showing that the silver could be the most suitable for coupling with PVDF-TrFe. Depending on the electrode shape, the piezoelectric voltage decreased with the decrease of the bottom and top electrodes overlapping area and this is common trend for all three studied metals. Thus, at maximal mass load of 100 g and bending frequency of 50 Hz, it can be observed that the silver rectangular electrodes could produce piezoelectric voltage (U_{PE}) of 916 mV, the meander electrodes could give 840 mV and the side comb electrodes – 446 mV. In comparison to the silver, the gold is the next possible solution, due to the very close results in term of piezoelectric voltage – 800, 768 and 362 mV - produced from rectangular, meander and side comb shaped electrodes, respectively. Although the technological compatibility of the aluminium, which is preferable electrode in most of the microelectronic device, here its weak piezoelectric response and high sensitivity to the electrodes shape make it not preferable for interfacing with PVDF-TrFe.

The shape of the generated piezoelectric signal from PEN/Ag/PVDF-TrFe/Ag sample at relatively small load of 80 g with frequency of 50 Hz is shown in figure 3 as an example for strong signal, although the distortions presence. It could be seen that the maximal voltage at this relatively low bending intensity was 664 mV (effective value rms was 265 mV), which is sufficient to drive some specialized power management circuits for accumulation of charge, such as ADP5090 ultralow power boost regulator that can be used for low power consumer supply [16].

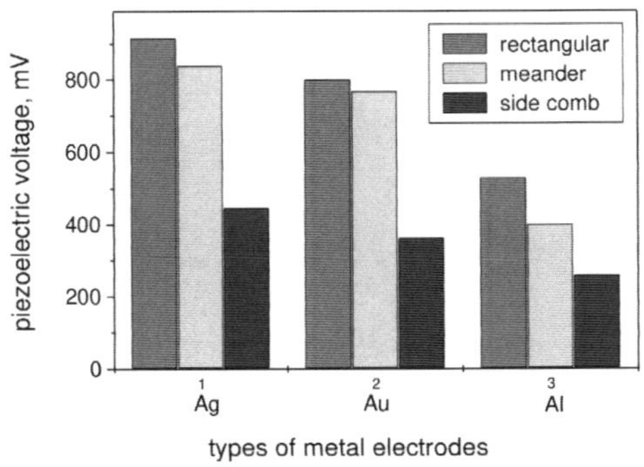

Figure 2. Comparison of the metals and their pattern suitable for PVDF-TrFe piezoelectric flexible elements at maximum mass load of 100 g and 50 Hz.

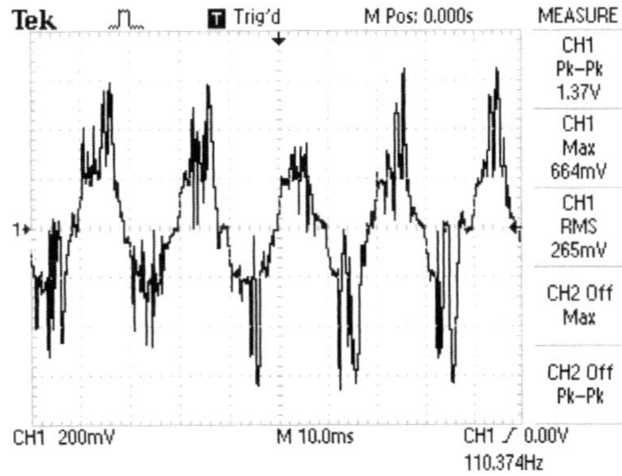

Figure 3. Piezoelectric voltage generated from PEN/Ag/PVDF-TrFe/Ag sample with rectangular electrodes at mass load of 80 g and frequency of 50 Hz.

For the determination of optimal performance of all nine produced combinations, the long term stability should be also considered. The results are presented in figure 4 a-c. When the samples are shacked and tested after 1500 repeating bending cycles, the speed of degradation is strongly dependent on the electrode shape. As can be seen from figure 4, the side comb configuration exhibited the most stable behaviour with variation of the piezoelectric voltage (not greater than 1.9 % from its initial value).

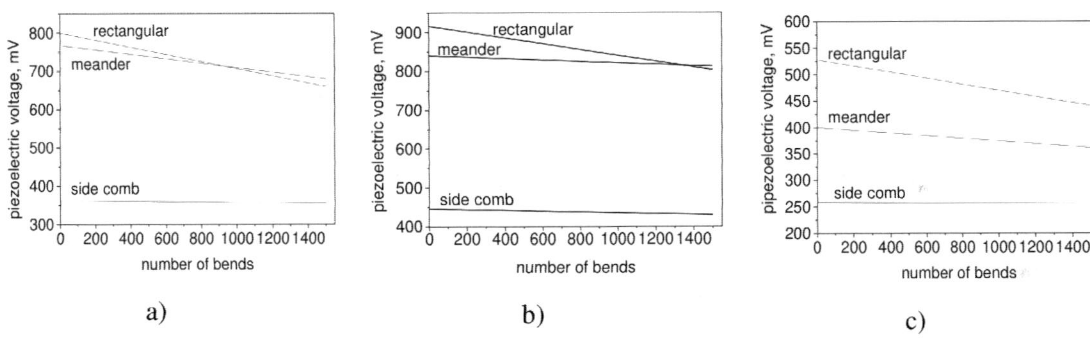

Figure 4. Stability of the piezoelectric voltage produced from PVDF-TrFe samples with different metal electrodes and different patterns of the electrodes at maximum mass load of 100 g and low frequency of 20 Hz: a) silver; b) gold; c) aluminium.

Electrical stability of the samples is strongly dependent on the wetting ability of the different electrode materials from the ink and the screen printed polymer coating morphology. As could be seen from the microscopy images in figure 5, the density of the PVDF-TrFe coating is relatively low, as compared to coating with the same thickness printed on silver or aluminium. This can explain the greatest instability of the piezoelectric voltage ($\Delta U_{PE\ @1500\ bends}$) generated from the samples with gold electrodes after 1500 bends, despite of the pattern. The sample with gold rectangular electrode exhibited 17.5 % decrease in the piezoelectric voltage, the meander shaped gold electrodes showed 11.4 % decrease, and the side comb - 1.9 %. For comparison, the same data about the aluminium are as follows: 16.4 %, 9.5 % and 1.7 % respectively. The samples with silver electrodes demonstrated the most stable behaviour with decrease in the piezoelectric voltage, respectively 12.3 % for rectangular electrode, 8.6 % for meander shape and 1.5 % for side comb pattern. As a general trend, the samples with side comb structures are characterized with the greatest stability of the piezoelectric signal after multiple bending, due to the smallest contact zones between electrode and polymer films. However, due to the poor coverage, the collected charge was the lowest possible from the three available patterns and the produced voltage barely reached amplitude beyond 480 mV. On the other hand the high density rectangular electrodes ensured high piezoelectric voltage up to 1 V, but it gradually decreased almost twice at the end of the test. The meander shaped electrodes accept intermediate position in the cases of all three studied metals with regards of piezoelectric yield and durability.

As can be seen from figure 5a the polymer printed on gold electrode is characterized with relatively low density coating, figure 5b shows finer granular and more homogenous structure of the PVDF-TrFe with higher density when printed on aluminium electrode and the coating in figure 5c, which is printed on silver exhibits the highest regularity and density. Although some point defects could be noted, the film in between them is smooth. This is a precondition for stable and uniform contact formation at the interface between the piezoelectric polymer and metal electrode with absence of strain effects which tent to peal of the polymer and corrupt the electrical connection between the coatings, resulting in piezoelectric voltage instability.

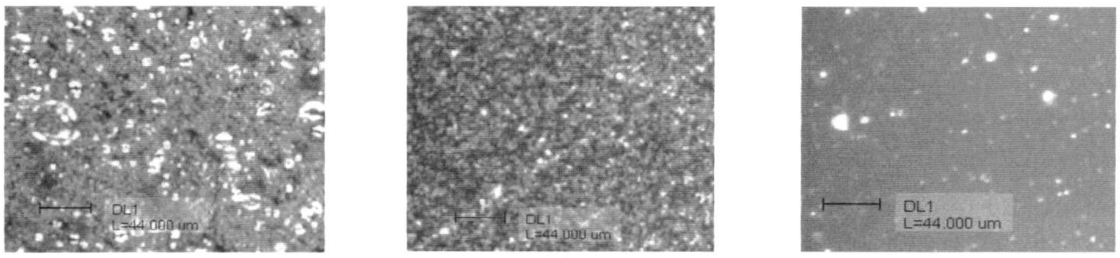

a) b) c)

Figure 5. Microscopic images of PVDF-TrFe screen printed on a) gold, b) aluminium and c) silver electrodes.

In all cases of patterns and variety of metals with different electrical conductivity, the overall conductivity of the samples was limited by the piezoelectric material and was in the range between 7 and 24 nA at the studied low mass loading and frequency of bending. Therefore, working as energy harvesting, these elements are able to produce electrical power P in the range of 6.3 μW and 22.8 μW which could be successfully accumulated in the micro-supercapacitors specially designed for work with high impedance energy harvesting systems.

Considering the relation between the piezoelectric coefficient g_{31}, the voltage generated V when stain σ is applied to piezoelectric film with thickness t ($g_{31}=V/\sigma.t$) it could be calculated that the maximum possible coefficient is 0.332 V.m/N when silver electrode is implemented in the harvesting design and give 930 mV at 100 g/cm^2 (0.098 N/m^2) over PVDF-TrFe coating with thickness of 3 μm. To the best of our knowledge, this value has not been previously reported and it is superior to this one of commercial thick film convertors which highest value is 0.06 V.m/N [17]. This result can be ascribed to the well-arranged high density polymer coating onto silver electrode, which form uniform and adherent interface at bending.

Based on the above achieved results the following comparison of the piezoelectric energy harvesting performance could be made (Table 1).

Table 1. Comparison of the basic properties of piezoelectric harvester with three types of metal electrodes and patterns.

	Ag electrode			**Au electrode**			**Al electrode**		
	rectang.	meand.	s.comb	rectang.	meand.	s.comb	rectang.	meand.	s.comb
U_{PE}, mV	916	840	446	800	768	362	528	400	258
ΔU_{PE} @1500 bends, %	12.3	8.6	1.5	17.5	11.4	1.9	16.4	9.5	1.7
Polymer microstru-cture	Smooth, uniform and dense coating			Relatively low density coating			Fine granular, homogeneous		
P, μW	~22.8			~13			~6.3		

IOP Conf. Series: Materials Science and Engineering **876** (2020) 012006 doi:10.1088/1757-899X/876/1/012006

4. Conclusions

Flexible polymer based energy harvesting elements were produced by simple processing and great compatibility with variety of substrates and electrode materials. They were investigated in terms of electrode effect on the piezoelectric behavior, according to the metal type and electrode patterns. It was found that the silver electrodes are more favorable in terms of produced piezoelectric voltage and quality of the printed polymer coating on them. It was also demonstrated that the side comb shape electrodes provide the highest stability of the piezoelectric voltage after great number of repeating bending cycles, despite of the electrode material nature. As a general conclusion, the meander electrodes could be a balanced compromise between electrical performance and mechanical stability. Another key outcome of the experimental work is the obtaining of a well controllable screen printed coating on surfaces with different wettability with new piezoelectric polymer ink, based on P(VDF-TrFE) material. Future work will be focused on experimental study of the frequency and temperature dependences of the dielectric permittivity and loss factor of flexible energy harvesting with screen printed PVDF-TrFe films with the proposed electrodes topologies.

5. References

[1] Yazid E and Nugraha A S 2019 *Int. J. Mech. Eng. Robot. Res.* **8** 940.
[2] Liu H, Zhong J, Lee C, Lee S-W and Lin L 2018 *Appl. Phys. Rev.* **5** 041306.
[3] Safaei M, Sodano H and Anton S 2019 *Smart Mater. Struct.* **28** 113001.
[4] Surmenev R A, Orlova T, Chernozem R V, Ivanova A A and Surmeneva M A 2019 *Nano Energy* **62** 475.
[5] Ponnamma D, Parangusan H, Tanvir A and AlMa'adeed M A A 2019 *Mat. Design* **184** 108176.
[6] Eddiai A, Meddad M, Farhan R, Mazroui M and Guyomar D 2019 *Superlattice Microst.* **127** 20.
[7] Defebvin J, Barrau S, Lyskawa J, Woisel P and Lefebvre J-M 2017 *Compos. Sci. Technol.* **147** 16.
[8] Kim J, Ji J-H, Shin D-J, Yoon S and Koh J-H 2019 *Appl. Surf. Sci.* **494** 1000.
[9] Zhao X L, Wang J L, Liu B L, Tian B B, Zou Y H, Sun S, Sun J L, Meng X J and Chu J H 2014 *Appl. Phys. Lett.* **104** 2013.
[10] Patel S, Chauhan A and Vaish R 2014 *J. Appl. Phys.* **115** 1.
[11] Seo J, Son J Y and Kim W-H 2019 *Mat. Lett.* **238** 294.
[12] Asadi K, Leeuw D M de, Boer B de and Blom P W M 2008 *Nat. Mater.* **7** 547.
[13] Pabst O, Perelaer J, Beckert E, Schubert U S, Eberhardt R and Tünnermann A 2013 *Org. Electron.* **14** 3423.
[14] Pi Z, Zhang J, Wen C, Zhang Z and Wu D 2014 *Nano Energy* **7** 33.
[15] Aleksandrova M, Kurtev N, Videkov V, Tzanova S and Schintke S 2015 *Microel. Eng.* **145** 112.
[16] https://www.analog.com/media/en/technical-documentation/user-guides/EVAL-ADP5090_UG-708.pdf
[17] Gusarova E 2015 "Flexible devices for energy harvesting based on printed organic piezoelectric P(VDFTrFE) materials" Université Grenoble Alpes.

Acknowledgments

This work was financially supported by Bulgarian National Science Fund, grant number DH07/13.

IOP Conf. Series: Materials Science and Engineering **876** (2020) 012007 doi:10.1088/1757-899X/876/1/012007

Design and Implementation of Interface Circuits Intended for Printed Piezoelectric Micropower Harvesters on Flexible Substrates

I Pandiev[1], M Aleksandrova[2], G Kolev[2]

[1]Department of Electronics, TU-Sofia, 8 Kl. Ohridski Blvd, 1000 Sofia, Bulgaria
[2]Department of Microelectronics, TU-Sofia, 8 Kl. Ohridski Blvd, 1000 Sofia, Bulgaria

ipandiev@tu-sofia.bg

Abstract. Interface circuits with low power dissipation is proposed and implemented, which is useful for efficient AC/DC voltage conversion of thin-film piezoelectric micro-power stack mounted harvesters. The focus is on low-power (< 1 μW) elements consisting of printed PVDF-TrFE piezoelectric polymer on $BaSrTiO_3$ coated flexible substrate with total thickness of the stack 3.1 μm. Using silver thin films as bottom and top electrodes, the samples exhibited stable piezoelectric rms voltage between 200 mV and 400 mV produced from functional area of ~1.5 cm^2 when stimulates it with sine-wave vibration with frequency of 50 Hz and intensity equivalent to mass loading between 1–80 g. The current registered from a single harvesting element is not greater than 1 μA. For this type of harvesters two power processing circuits intended are designed and built. The first circuit is a voltage doubler, for which the rectified output voltage is approximately equal to twice of the amplitude of the input voltage. The second circuit is a voltage quadrupler; as compared to the first one it provides higher voltage for charging a small and thin-size chip supercapacitor connected to the output port, but due to diodes voltage drops a smaller value of the energy efficiency can be obtained. For the implementation of the electronic circuits, low-power Schottky diodes with a forward voltage below 100 mV at current up to 0.1 mA are chosen. An experimental study to verify the efficiency of the proposed circuits is performed with laboratory made thin-film piezoelectric harvesters.

1. Introduction

Mechanical-to-electrical energy conversion for application as alternative energy source is a process by which elastic wave causes deformation in some materials resulting in dipole formations, thus converted to electrical energy due to direct piezoelectric effect. As the sources of mechanical waves could strongly vary from human motion, cars or industrial machines vibrations, to weather conditions (wind flow, water fall) and etc., the application of such elements are in the bioelectronics, automotive and industrial electronics for supplying low-power and portable consumers [1]. The piezoelectric energy harvesting (PEH) elements have to be sensitive, effective, but in the same time small sized and light weighting, especially in the field of health care applications. This is achieved by using thin films of piezoelectric materials, which is in opposition with the required high efficiency that could be achieved from small volume of material. One of the issues toward commercialization of such battery-less and portable devices is the inability of the piezoelectric coatings to produce electrical power beyond few micro-Watts. Additionally, it is preferable to grow the entire element on flexible substrate

for greater compactness. The implementation of materials with naturally high piezoelectric coefficients partially solves the above highlighted problems [2]. They are usually oxide based in composition ($BaTiO_3$, $PbZrTiO_3$, $KNbO_3$ etc.) and are characterized with high Young modulus and brittleness [3]. Introduction of material with elastic coefficient closer to that of the flexible substrate neutralizes the stress that the oxide material experiences and would increase the stability of the generated signal due to the enhanced mechanical stability of the energy converting coating. Very often the piezoelectric polymers can successfully serve like such kind of buffer material [4].
The flexible harvesting elements reported in this paper are made from piezoelectric polymer poly(vinylidene fluoride-co-trifluoroethylene) P(VDF-TrFE) characterized with low Young modulus of 0.61 GPa [5] favourable for durability at bending and twisting substrates and piezoelectric ceramic barium strontium titanate ($BaSrTiO_3$ or BST) with high piezoelectric coefficient of 33.4 pC/N [6] preferable for strong piezoelectric response provoked even by weak dynamic load. The combination of both materials results in enhanced piezoelectric voltage generating abilities and long term stability of the signal at multiple repeating bends or twists.

For converting the mechanical strain energy into electricity by a PEH, the basic ac/dc converter is as a diode rectifier is connected to a rechargeable battery or a supercapacitor. The main disadvantage of this type of interface circuits is that as a result of the voltage amplitude fluctuations, generated by the PEH, the power that is extracted can be significantly reduced [7]. For batteries and supercapacitors, a fixed value of the voltage is required. For the last few years the analysis of the literature resources shows a wide variety of power energy-harvesting circuits, based on active electronic components [8-12]. In this work the focus is on the interface circuits in [8], [10] and [12] that operate at an input voltage with rms value less than 1 V and the output power is below 10 μW, because the fabricated printed PEHs on flexible substrate produce an ac voltage with rms value up to several hundreds of millivolts in contrast with the most wide spread processing circuits dealing with light harvesters that are able to produce dc voltage of few volts [13].

The goal of this paper is to present the design process and experimental testing of interface circuits that have to be connected between flexible P(VDF-TrFE)/BST – based piezoelectric generators and a micro supercapacitor, used as a load. Furthermore, the developed interface circuits are based on voltage doubler rectifier which has been initially presented for this type of input source in [8,13]. The proposed circuit is intended to use for micro-power (1–10 μW) thin-film PEHs and overcomes the some of the aforementioned limitations of the existing power processing circuits.

2. Structure and Preparation of Printed Piezoelectric Micropower Harvesters on Flexible Substrates

The flexible substrate was thermal resistant polyethylene naphthalate (PEN). Silver bottom and top electrodes with thickness of 450 nm were produced by thermal evaporation in vacuum at base pressure 10^{-6} Torr. Conventional for the MEMS technology RF sputtering was used for the BST film at argon pressure of $2.5x10^{-2}$ Torr without additional oxidation. Sputtering voltage of 0.65 kV was set for 50 min at plasma current of 150 mA. Thus low temperature deposition process was established with plasma power of 97.5 W. The film thickness was 560 nm measured by Tencor α-step profilometer.

For the P(VDF-TrFE) ink low-cost screen printing technology was used. Particularly, the screen printing is well known and preferable additive manufacturing method, because of the simple and fast process, easy control over the deposition, as well as due to the possibility for uniform, high density coatings deposition on large area with no matter of the substrate type [14]. In addition, desired patterning of the films can be achieved in situ during the deposition without consequent photolithographic process (Fig. 1a). In this way, the proposed flexible energy harvesting is characterized with satisfactory efficiency, stability and price. After the deposition, all samples were annealed at 100°C by recommendation of the manufacturer for removing the residual solvent and initiate crystallization of the piezoelectric phase in the polymer coating. The temperature treatment is the reason for selection of BST as piezoelectric oxide from the great variety of modern lead-free piezoelectric materials, because annealing at this temperature wouldn't cause phase transition process and properties (permittivity and loss factor) variation in the barium strontium titanate [15]. The thickness of the fabricated polymer coating after annealing was 1.6 μm. To avoid penetration of the silver paste's solvent deeper in the polymeric coating, the silver top electrode was not fabricated by screen printing of silver paste, but again vacuum thermal evaporation was used through shadow mask for segmenting the coating (Fig. 1b).

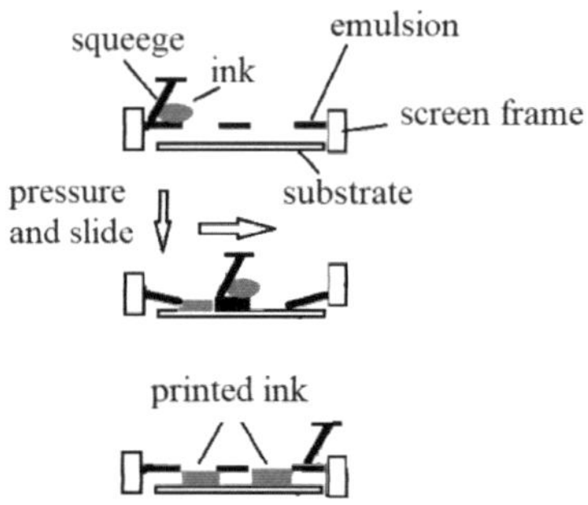

Figure 1a Principle of screen printing of inks

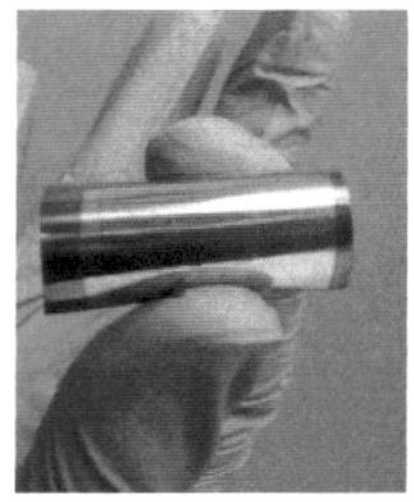

Figure 1b Photo of the prepared sample

3. Electrical circuits and theoretical analysis

Interface circuits based on passive doubler (VD) rectifiers are proposed for converting the energy from the presented micro-power piezoelectric generators. In addition to straightening, these rectifiers also result in increased voltage by appropriately connecting a group of diodes and capacitors. Theoretically, the amplitude of the input voltage increases integer times; taking into account the voltage drops on the diodes. In order to avoid a significant decrease in the output voltage, Schottky diodes are chosen to use in the proposed circuits instead of using rectifying silicon diodes. By using a supercapacitor as an external load to the VD rectifiers, it can hold little energy at low voltage and is used as energy storage element for a short time.

3.1. Voltage doubler circuit

In Fig. 2 is a circuit diagram of a rectifier with voltage multiplication factor of two (voltage doubler circuit) is given. For this circuit, the operational principle is as follows. During the first half wave of the input signal (the polarity of the input voltage is indicated by characters without brackets) capacitor C_1 is charged through the diode D_1 to the value, equal to the amplitude of the input voltage – $V_{C1} = V_{p0}$. In the next half wave, the input signal changes its polarity (the characters in brackets). The diode D_1 is turn off, on it is applied reverse voltage equal to the sum of the voltage of the capacitor C_1 and the input voltage. The open-circuit voltage drop (in condition that there is no load – without C_2 and the internal resistance r_p is neglected) over the piezoelectric capacitance C_p is given by [16]

$$V_{p0} = I_p / \omega C_p . \tag{1}$$

The maximum average power extracted from a PEH occurs when the VD is one-half of the peak voltage is [7]

$$P_{PEH,av} = V_{p0}^2 \omega C_p / 2\pi \tag{2}$$

at $V_B = 0.5 V_{p0}$ and $P_{PEH,av}$ is the extracted power and subscript *av* denotes its average value.

The no load peak value of the output voltage, is found by

$$V_B = 2 I_p / \omega C_p . \tag{3}$$

The internal capacitance and resistance of the PEH can be used as a part of the voltage doubler circuit, shown in Fig. 2. The PEH equivalent circuit is a simplified model that represents the dominant components of the generated current and output impedance ($Z_p = r_p \parallel C_p$) of a typical thin film low-power harvester. When a capacitor C_2 is connected to the VD, the output voltage becomes a rippled DC voltage, and $V_B \le 2V_{p0}$. During this time interval, the diode D_2 is turn on and the capacitor C_2 is charged through it and through the capacitor C_1, which transfers a part of its charge up to a voltage equal to the sum of the amplitude of the input voltage and the voltage of C_1. Furthermore, the voltage drop over the diode D_2 limits the value of the produced piezoelectric voltage (i.e., $v_p(t) \le V_B \le 2V_{p0}$).

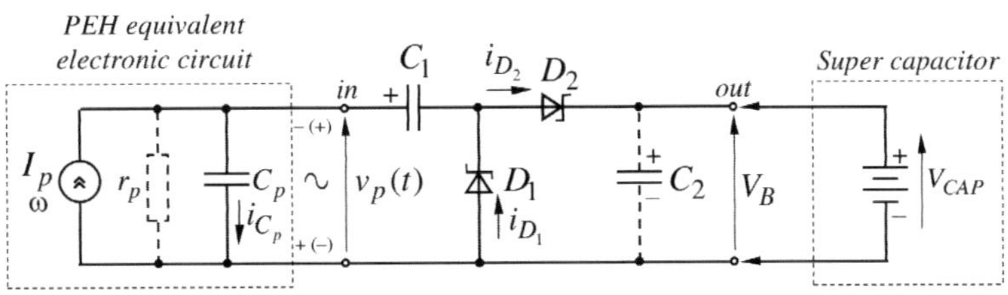

Figure 2 Circuit diagram of voltage doubler circuit for charging supercapacitor.

In comparison with the full-wave rectifiers [7, 12] the circuit in Fig. 2 has only one diode forward voltage drop in the current path at any one period of the input voltage, which increases the V_{out} and further reduces the dissipation power in comparison.

3.2. Voltage quadrupler circuit

In Fig. 3 is a circuit diagram of a rectifier with voltage multiplication factor of four (voltage quadrupler circuit) is given. This circuit consists of four capacitors and four diodes. Moreover, this circuit can be considered as a complicated variant of the voltage doubler. Compared to it, a lower value for energy efficiency can be obtained, but a higher value of the output voltage can be achieved.

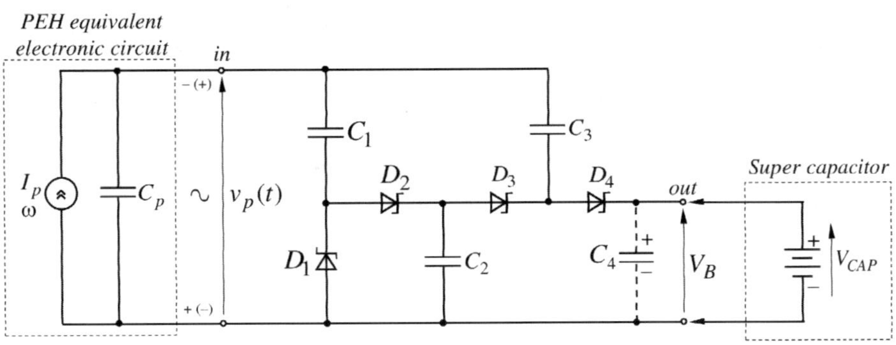

Figure 3 Circuit diagram of voltage quadrupler circuit for charging supercapacitor.

The operational principle of the circuit in Fig. 3 is analogous to the voltage doubler. In the concrete case, for the third half-wave voltage (again the polarity without brackets), the diode D_2 conducts and the capacitor C_1 is recharged to voltage $V_{C1} = V_{p0}$. The capacitor C_3 is charged through the diode circuit D_3, the capacitor C_2 and the capacitor C_1. The maximum voltage drop of capacitor C_3 is determined by

$$V_{C3} = V_{p0} + V_{C2} - V_{C1} = 2V_{p0}. \quad (4)$$

During this time interval the diode D_2 is off and the reverse voltage across it is equal to the voltage over the capacitors C_2 and C_3, which through the conductive diodes D_1 and D_3, are connected in parallel to it, i.e. $V_{D_2\,\mathrm{Rev}} = 2V_{p0}$. The load capacitor C_4 is connected between the cathode of diode D_4 and one terminal of the PEH. Since the total number of the capacitors is four, the maximum value of the output rectified voltage should be four times greater than the voltage V_{p0}.

For both circuit variants, the capacitors are selected such that the fluctuations of the output voltage do not exceed 5% [7] of the nominal DC voltage for a specific load. To achieve greater unification and manufacturability, the capacitors can be selected with the equal capacitance, which is of average value and guarantees up to 5% ripples of the rectified voltage [17]

$$C = I_B (fU_B)^{-1} 2(k+2) \times 10^6, \ \mu F \ \text{at} \ k = 4. \quad (5)$$

The rectifier circuits in Fig. 2 and 3 have relatively large value of the internal resistance and a falling output characteristic, so they are used when exactly such a characteristic is required for a specified load or when the load is for a small current with constant value. By using a supercapacitor as an external load, it can hold little energy at low voltage and is used as energy storage element for a short time.

4. Experimental verification

Both interface circuits have been implemented on prototyping circuit boards using commercially available discrete components. The used Schottky diodes are type PMEG3002TV and 1N5817, and the capacitors are multilayer ceramic components with 1 μF capacitance. The designed circuits are intended for charging energy storage elements such as low-power thin-size supercapacitor. For the purpose of this work we used a micro supercapacitor, type XH311HU-IV07E (from Seiko Instruments) with initial capacitance more than 0.022 F and nominal capacity 12 μAh. The shaker is specially built for those specific types of PEHs and has been used to get the desired input vibrations equivalent to mass load up to 100 g and low frequency up to 50 Hz. A power signal generator Г3-109 is used to apply the desired ac voltage to the shaker electrical circuit, realizing cantilever type mesurement. To measure the voltage drop over the supercapacitors, an Agilent 34405A 5 ½ digit multimeter and a VC820 4 ½ digital multimeter with an opto-electronic interface to personal computer are used.

To illustrate the operational principle of the voltage doubler and quadrupler circuits the transient waveforms are shown in Fig. 4 and Fig. 5. The experimental study is performed with a connected load capacitor 1 μF at the output. For the both circuits the amplitude of the input signal is 0.5 V (or 1 Vp-p) and the nominal frequency is 50 Hz. It's obtained by sine-wave generator SFG-1013. The measured forward voltage drop over the rectifier diodes is low and approximately equal to 0.1 V. As a result, the rectified output voltage is set to 900 mV – 30% of the capacity (Fig. 4) and 1.7 V – 50% of the capacity (Fig. 5), respectively. Those results guarantee the applicability of the used electronic circuits for low voltages, generated by thin-film PEHs.

After verification of the operational principle of the two circuits, a direct charging of the 3.3 V micro supercapacitor type XH311HU-IV07E from a single PEH at two levels of excitation is performed. The obtained maximum peak-to-peak output voltage swing of the input signal, generated by single PEH, is approximately equal to 300 mVp-p and the frequency is up to 50 Hz. Fig. 6 and 7 shows the plots of an experimentally measured voltage versus charging time from a single thin-film printed piezoelectric micropower harvester by using the circuits in Fig 2 and Fig. 3. Fig. 6 shows the charge of the supercapacitor to 85 mV. As it can be seen, the charging time is approximately three hours. The charge rate of the capacitor in the initial time intervals is high and then gradually decreases to the level of the rectified output voltage. Fig. 7 also shows the charge of the supercapacitor but up to value equal to 160 mV. In this case, the charge of the supercapacitor is a little more than 6 hours, and at the initial time intervals, the charging rate is again with higher value.

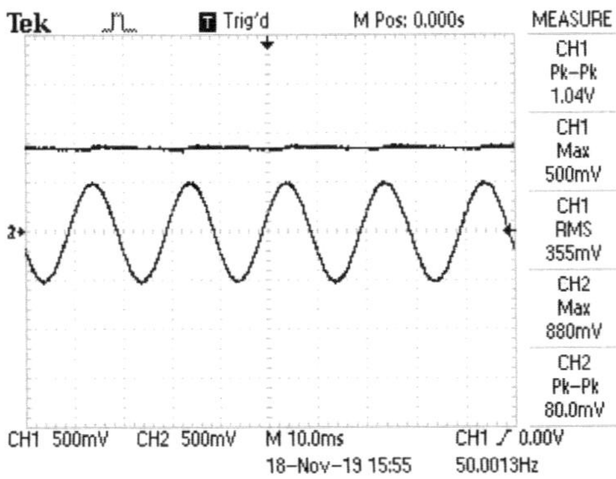

Figure 4. Measured transient waveforms for input and output nodes in the voltage doubler at 1 μF load capacitor.

Horizontal scale is *10ms/div*
Vertical scale is *500mV/div* – Channel 1 and 2

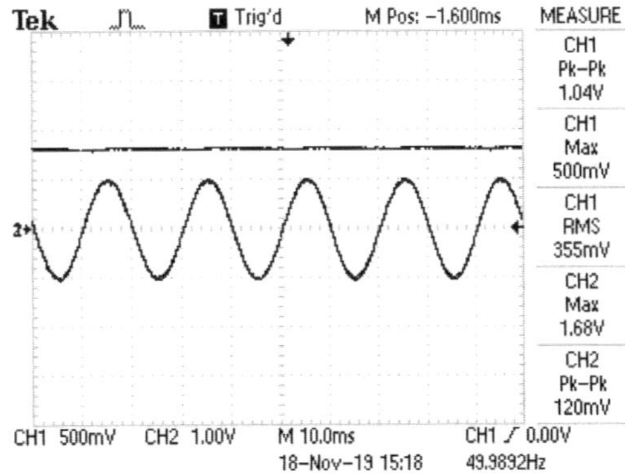

Figure 5. Measured transient waveforms for input and output nodes in the voltage quadrupler at 1 μF load capacitor.

Horizontal scale is *10ms/div*
Vertical scale is *500mV/div* – Channel 1 and *1V/div* – Channel 2

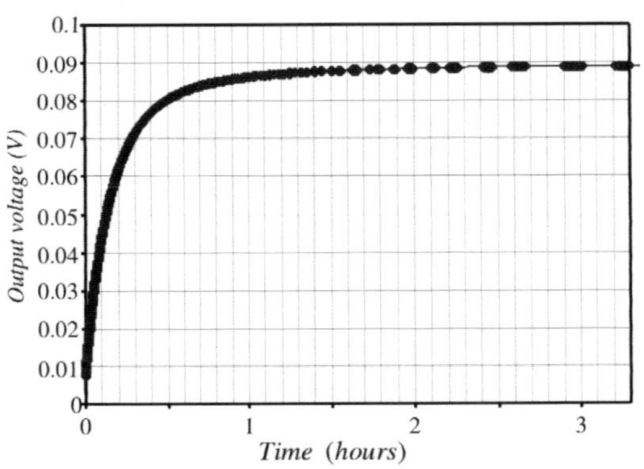

Figure 6. Experimentally measured voltage from single PEH on a micro supercarpacitor type XH311HU for voltage doubler circuit.

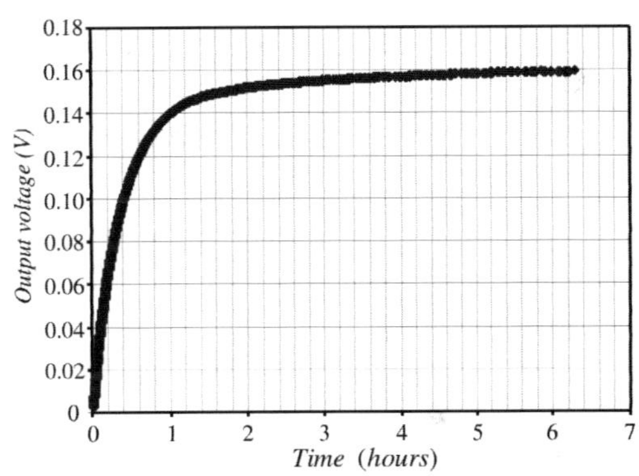

Figure 7. Experimentally measured voltage from single PEH on a micro supercapacitor type XH311HU for voltage quadrupler circuit.

5. Conclusion

In this paper, the behaviour of two conversion circuits based on the voltage doubler rectifier is evaluated for charging micro supercapacitors. The developed rectifier circuits are intended for working with mounted in stack thin-film PEHs, consisting of printed PVDF-TrFE piezoelectric polymer on $BaSrTiO_3$ coated flexible substrate with total films thickness of the piezoelectric coatings in single element equal to 3.1 μm and active piezotransducing area equal to 1.5 cm^2. The conversion circuit of voltage doubler rectifier has only one diode forward voltage drop in the current path at any one period of the input voltage, which provides slightly higher value of the energy efficiency. Its convenient for low power single PEHs. For the voltage quadrupler circuit lower value for energy efficiency can be obtained, but a higher value of the output voltage can be achieved. The voltage quadrupler can be used for mounted in stack thin-film PEHs.

The measurement results show the contribution of single PEH from the stack for charging supercapacitors. Due to the high internal impedance of the synthesized PEHs, conventional active full-wave synchronous rectifiers do not provide efficient power conversion. The maximum average power extracted from the single PEH is up to 1 μW. To overcome the low output voltage, Schottky diodes with a small forward voltage (≈ 100 mV) are chosen.

6. References

[1] Huicong L, Zhong J, Lee Ch, Lee S-W and Lin L 2018 *Appl. Phys. Rev.* **5** 041306
[2] Yang Z, Zhou Sh, Zu J and Inman D 2018 *Joule* **2** (4), p 642
[3] Villafuerte-Castrejón M E, Morán E, Reyes-Montero A, Vivar-Ocampo R, Peña-Jiménez J-A, Rea-López S-O and Pardo L 2016 *Materials* **9** (21)
[4] Mishra S, Unnikrishnan L, Nayak S K and Mohanty S *Macromol. Mater. Eng*, 2019 **304**, 1800463
[5] Information on http://www.nanopaint-tech.com
[6] Kwonet S R al *Smart Mater. Struct*. 2013 **22**, 115017
[7] Tabesh A. and Frechette L G *IEEE Trans. Ind. Electron*. 2010 **57** (3) p 840
[8] Du Sh, Jia Y, Zhao Ch, Amaratunga G, and Seshia A *IEEE Trans. on Ind. Electron.* 2018, **65** (9) p 7095
[9] Liang J, Zhao Y, and Zhao K *IEEE Trans. on Power Electron.* 2019 **34** (1) p 275
[10] Shareef A, Ling Goh W, Narasimalu S, and Gao Y *IEEE Trans. on Power Electron.* 2019 **34** (2) p 1446.
[11] Costanzo L, Lo Schiavo A, and Vitelli M *IEEE Trans. on Circ. and Syst.–I: Regular Papers*, 2019 **66** (3) p 1291
[12] Szarka GD, Stark B H and Burrow S G *IEEE Trans. Power Electron*. 2012 **27** (2), p 803
[13] Mahmoud H H, Hafiz A H, Fathy A and Abdellatif S O 2019 *Int. J. Electric. Electron. Eng. Telecommun*. **8**, p. 177.
[14] Chang W-Y, Fang T-H, Lin H-J, Shen Y-T and Lin Y-C *Journal of display technology* 2009 **5** p 178
[15] Yu Y, Zou H, Cao Q-F, Wang X-S, Li Y-X and Yao X 2015 *Ferroelectrics* **487**, p 77
[16] Dicken J, Mitcheson P, Stoianov I, and Yeatman E *IEEE Trans. on Power Electron*. 2012, **27** (11) p 4514
[17] Stefanov N 2002 Special rectifiers *Power supply devices*, (Sofia: Tehnika) chapter 4

Acknowledgments

The authors would like to thank Bulgarian National Science Fund, grant DH 07/13 for the financial support. Also, the authors would like to thank Prof. M. Antchev of the Technical University of Sofia for his support.

IOP Conf. Series: Materials Science and Engineering **876** (2020) 012008 doi:10.1088/1757-899X/876/1/012008

Biomechanics of fish swimming in a hydrokinetic turbine wake

M M Macias[1], L Lafuente[2] and A C P Brasil Junior[1]

[1]Mechanical Engineering Department, Faculty of Tecnology, Universidade de Brasilia, 70900190 Campus Darcy Riberio, Brazil

[2]Mathematical Department, Faculty of Engineering, Universidad de Cádiz, 11519 Puerto Real, Spain

marianela@unb.br

Abstract. In the present paper, the assessment of the hydrodynamics of the fish swimming in undisturbed and altered wake-flow is carried out identifying hydrodynamical forces and flow patterns of the fish swimming wake. URANS approach with k-ω/SST turbulence model are employed combining fish and turbine in the same simulation. Fish motion is realized using dynamically adaptive mesh. The actuator line method is employed to induce the wake of a hydrokinetic turbine, which is a simplified method that requires lower computational cost than full geometry simulations. This work brings a new numerical approach involving fish and turbine wake highlighting that fish swimming in the wake presents higher thrust forces than in the undisturbed flow, due to the x-component to velocity in the wake is lower than free flow velocity.

1. Introduction

Nowadays, hydrokinetic energy appears how a technological alternative in the context of sustainable energy. Based on the same principles as wind energy, hydrokinetic devices can produce electricity extracting kinetic energy contained in rivers, estuaries and, channels. Hydrokinetic turbine projects are growing considerably in the last decade however this technology is considered in the pre-commercial stage [1]. Environmental impact studies are a relevant issue yet in development. In literature are described main causes affecting aquatic life due to hydrokinetic devices installation such as sediment suspension, alteration of river hydraulic regimes, noise, electromagnetic fields, fish collisions with blades turbine, toxicity of paints and other chemicals and the installation of several hydrokinetic devices in arrangement [2,3]. Nevertheless, there are not found studies in the literature about hydrodynamic interaction between fish and turbine.

The present paper aims to study fish swimming biomechanics under altered flow conditions similar to hydrokinetic wake topology. Firstly, a numerical methodology is developed to promote tuna fish (*Thunnus atlanticus*) model swimming in undisturbed flow promoting a reliable description of the vortex wake and the computation of hydrodynamic forces. Secondly, the methodology established is employed to carry out analysis of tuna fish (under the same kinematic parameters) swimming in the wake of a hydrokinetic turbine. Transient forces and wakes are computed and compare in both cases, without and with turbine. Simulations were realized in the open-source software OpenFOAM® using the dynamical mesh adaptive tool to simulate fish motion and in the framework of the turbulence, were employed a URANS approach and the k-ω/SST closure turbulence model. Finally, the turbine wake is induced using actuator line model simulations.

2. Numerical methodology

2.1. Actuator Line Method

The Actuator Line Method [4] is a simplified methodology combining Navier-Stokes equation with the Blade Element momentum method (BEMM) to compute the flow around a turbine, calculating wake-flow and hydrodynamical forces on the blades. ALM not employs rotor geometry, blades are considered lines discretized as points and each point represents a blade section. Hydrodynamic forces are computed from an iterative process where the flow solved by Navier-Stokes equations provided information about attack angle and velocity of the flow which are employed in BEM method calculating lift and drag forces on blades (inputting 2D lift and drag forces and airfoil geometry). After all, forces are projected on the background Cartesian grid as body-force field using a three-dimensional Gaussian function and incorporated in the Navier-Stokes. This method is simulated from OpenFOAM® using the library *turbinesFOAM* [5].

In this work, the hydrokinetic turbine is a horizontal axis turbine HK10 designed by the Energy and Environmental laboratory at the University of Brasilia in the context of the P&D project AES Tiete HYDROK. The rotor has 4 blades and 2.2 m of diameter, blade sections are NACA4412 profiles. The operating point employed in the simulations is defined by the tip tip speed ratio, $TSR = \omega R/U$, and the free flow velocity, $U = 2.5 m/s$ and, rotation velocity, $\omega = 35 rpm$. For more information about the ALM simulations see [6] where a validation case is presented. The ALM reproduces the three-dimensional structures expected in a wake turbine and therefore, we use this simplified method to induce the wake-flow for the fish swims in the altered flow.

2.2. Fish geometry, kinematics and force computation

Tuna fish geometry is the same as analysed by [7]. The fish body and caudal fin were generated from elliptical and airfoil cross-sections and envelope equations provided by the referenced study, see Fig. 2. Fish swims using the body/caudal fin (BCF) description under carangiform mode. Movement is characterized by undulatory waving motion of its flexible body from head to fin tail that generates a propulsive thrust force and it is described by sinusoidal equation $h(x,t)$ [8] with variable amplitude $a(x)$ such as

$$h(x,t) = a(x)\sin(k\,x - \omega\,t) \tag{1}$$

$$a(x) = a_0 + a_1 x + a_2 x^2 \tag{2}$$

where $h(x,t)$ is the midline fish displacement, ω tail-beat frequency and *k* the wake number, being $k = 2\pi/\lambda$ the wavelength of propulsive traveling wave. The coefficients a_0 , a_1 and a_2 are determined experimentally and reported in literature. In this work, the kinematic parameters employed were defined from the study of [7] being $\lambda = 1.675$, $\omega = 13\, rad/s$ and the coefficients $a_0 = 0.0$, $a_1 = 0.14$ and $a_2 = 0.000236 m^{-1}$. Simulations are carried out considering the fish steady in a constant velocity flow.

Hydrodynamic force acting on fish in *i*-direction is computed from the instantaneous velocity and pressure fields, using an integral of the stress-tensor over the fish surface [9]. It is given by

$$F_i(t) = \int_S (-p n_i + \tau_{ij} n_j) dS \tag{3}$$

where τ_{ij} is the stress-tensor components and n_j is the normal of the fish surface. The force time-averaged value (F_i) is obtained integrating the instantaneous force coefficient in an entire cycle of fish swim. If $F_i = 0$ (*x* is flow direction) the fish swims in a conditions of dynamical equilibrium between the drag and thrust forces. For positive values the fish accelerates upstream (thrust is greater than drag) and for negative value the effect is opposite.

IOP Conf. Series: Materials Science and Engineering **876** (2020) 012008 doi:10.1088/1757-899X/876/1/012008

2.3. Computational domain, numerical mesh and boundary conditions

The computational domain employed in the simulations is a rectangular prism with dimensions 4D×4D×7D. The inlet flow surface is placed 2D upstream of the turbine and the fish is 2D downstream to rotor (placed at system origin). In figure 3, the rotor is represented by a red rectangle just to help in understanding. Boundary condition imposed on inlet surface is uniform velocity and 5%turbulence intensity. On the outlet surface, a reference pressure is considered to open flow conditions and the lateral walls are free-slip surfaces. Fish surfaces are assumed to be non-slip boundaries, which move according to equations 2 and 3. Due to in ALM simulations there is no turbine surface, no-slip wall boundary condition has not to be imposed.

Numerical mesh was created in OpenFOAM® using the tool *snappyHexMesh*. The mesh on fish surface was highly refined with multiple layers of prismatic elements, to properly simulate of the boundary layer gradients and to maintain the parameter y^+ close to one, as required by the $k - \omega$/SST turbulence model [10]. A mesh convergence study was carried out to guarantee that results do not depend on mesh discretization for fish swimming in undisturbed flow. For the turbine, a mesh convergence was realized in the work [6]. So, in the simulation combining rotor and fish, the element size was kept ensuring the mesh reliability and the number of elements is close to 4 million.

The time-step used in simulations was 10^{-3} keeping the Courant number minor to the unit. For all simulations, free flow velocity was 0.7m/s and tail beat fish frequency 13 rad/s. The rotor kept the Tip speed ratio value was kept to nominal operating condition of the rotor (TSR=1.6), being free flow velocity 0.7m/s and rotational velocity 1.01 rpm.

Figure 1. Tuna fish 3D geometry.

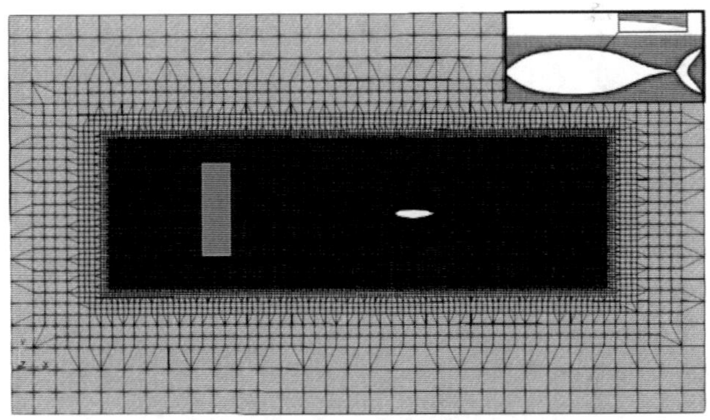

Figure 2. Numerical mesh.

3. Results

Fish wake structure is a relevant issue to understand the resultant propulsion force, evaluating the drag or thrust effect in the framework of the wake topology. In figure 3, dimensionless x-velocity component field and z-vorticity visualizations allow the comprehension of the fish wake swimming in undisturbed flow and its periodic vortex emission (like a reverse von Kármán vortices). In this situation, the fish is self-propelled and it moves backward flow experiencing a thrust force ($\overline{F_x} > 0$). Tuna fish presents a jet flow downstream which boosts the fish propulsion, with a vortex sequence organized like positive vortex (counter clockwise - in red colour) above fish symmetry line and negative vortex (clockwise - in blue colour) below that. Time history, for both cases, fish swimming in undisturbed flow and behind the rotor, are presented in figure 4. In both simulations, free flow velocity and fish tail beat frequency are kept equals to understand the effect of the flow turbine on a fish.

From figure 4, it is possible to see that signals begin to differ at time 7s. Around 8.8 s, the black signal corresponding to fish force swimming in the wake-flow achieve maximum values and then, it decreases remaining stable but without returning to the initial values.

Figure 3. Non-dimensional velocity and vorticity fields, (U/U_{∞}) and $(\omega L/U_{\infty})$ for fish swimming in undisturbed flow.

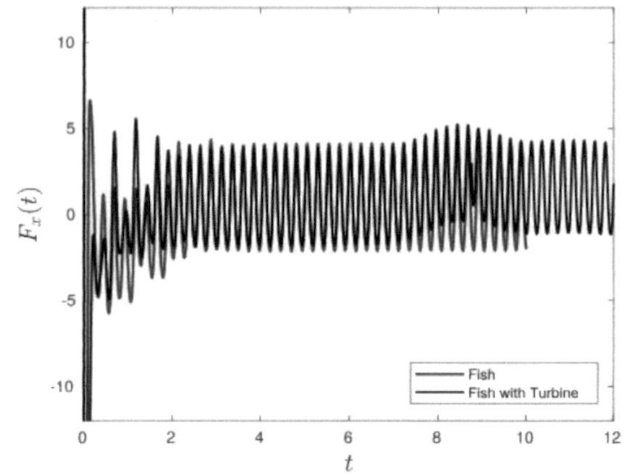

Figure 4. Time history of x-direction forces to fish swimming without and with turbine.

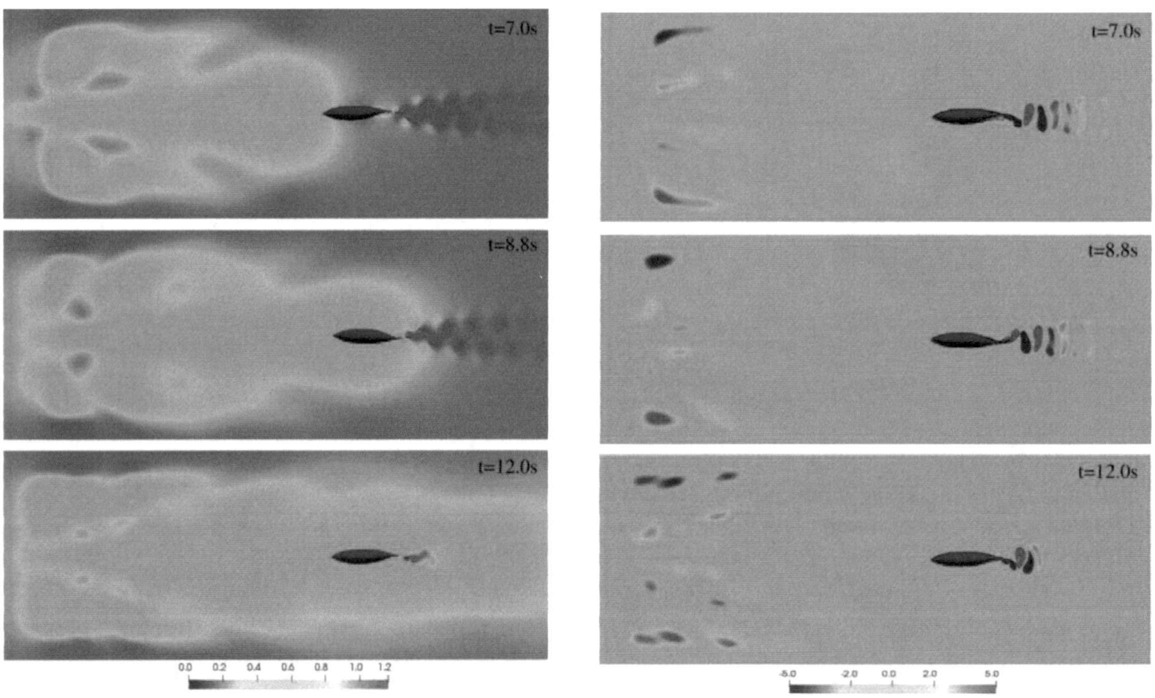

(a) (b)

Figure 5. Fish swimming in turbine wake at three different times. (a) Non-dimensional velocity (U/U_{∞}) and (b) vorticity fields, $(\omega L/U_{\infty})$.

From equation 3, mean force in x-direction is computed employing 4 periods, at 4s (not turbine interference) and 10s (last times simulated). Results achieving are $\overline{F_x} = 0.63N$ and $\overline{F_x} = 1.30N$, respectively. For all, fish experiences acceleration in opposite direction to the flow due to thrust-type force ($\overline{F_x} > 0$) .

From figure 5, we can be able to relates turbine wake evolution with time history in figure 4. Firstly, we verify that fish find the wake-flow at 7s. From that instant, fish swims at lower velocities and experiences higher forces due to tail beat frequency is kept constant. At instant 8.8 s, where maximum forces values were noted, a region of the fluid with even lower velocity reaches the fish. Last, at higher times, fish is swimming immersed in a permanent wake where force signal oscillates around a mean value. In figure 5, when t=12s, fish wake appears shorter due to it is pushed by the flow of wake turbine but the wake maintains a reversal von Kármán vortex street configuration generating propulsive force.

4. Conclusion

Numerical simulations for tuna fish swimming in an undisturbed flow and an altered flow were presented in this work. The altered flow has the hydrokinetic turbine wake topology and it was induced through simulations by the simplified actuator line method. Results obtained from CFD study allow us to conclude that to both cases (with and without turbine), fish experiences thrust force appearing the wake like a reverse von Kármán vortex street generating a jet flow downstream which boots the fish propulsion. But, fish swimming in the turbine wake (at the same tail beat frequency) presents a higher mean force due to the flow velocity is lower. Although, these results contribute to the understanding of fish swimming in altered flow. It should be necessary to compute the propulsive efficiency at the frequency equilibrium condition and to analyse the vortex effect to conclude fish performance behaviour.

5. References

[1] Laws N D and Epps B P 2016 Technology, research, and outlook *Renew. Sust. Energ.* **57** pp 1245–1259
[2] Kartezhnikova M and Ravens T M 2014 Hydraulic impacts of hydrokinetic devices *Renew. Energy* **66** pp 425–432
[3] Fallon D, Hartnett M, Olbert A and Nash S 2014 The effects of array configuration on the hydro-environmental impacts of tidal turbines *Renew. Energy* **64** 10–25
[4] Sørensen J N and Shen W Z 2002 Numerical Modeling of Wind Turbine Wakes *J. Fluids Eng.* **124** pp 393
[5] Yu Z, Zheng X and Ma Q, 2018 Study on Actuator Line Modeling of Two NREL 5-MW Wind Turbine Wakes *Appl. Sci.* **8** p. 43
[6] Macias M M, Mendes R C F, Brasil Junior A C P and Felamingo T O Hydrokinetic turbine simulations using actuator line method *Proc. of the COBEM 2019: 25th ABCM International Congress of Mechanical Engineering* Uberlandia, MG,Brasil ABCM
[7] Zhu Q, Wolfgang M J, Yue D K P and Triantafyllou M S 2002 Three-dimensional flow structures and vorticity control in fish-like swimming *J. Fluid Mech.* **468** pp 1-28
[8] Lighthill M J 1969 Hydromechanics of Aquatic Animal Propoulsion *Annu. Rev. Fluid Mech* **1** pp 413-446
[9] Borazjani I and Sotiropoulos F 2008 Numerical investigation of the hydrodynamics of carangiform swimming in the transitional and inertial flow regimes *J. Exp. Biol.* **211** pp 1541–1558
[10] Menter F R 1994 Two-equation eddy-viscosity turbulence models for engineering applications *AIAA Journal, American Institute of Aeronautics and Astronautics* **32** pp 1598-1605

Acknowledgments

This work was conducted during a scholarship supported by the International Cooperation Program of CAPES (Brazilian Federal Agency for Support at the University of Cádiz. Financed by CAPES within the Ministry of Education of Brazil. It was partially supported also by the grants of project SEFAC-UnB, in a framework of R&D project of Brazilian agency of Energy – ANEEL.

IOP Conf. Series: Materials Science and Engineering **876** (2020) 012009 doi:10.1088/1757-899X/876/1/012009

Comparison of experimental and model liquid distribution in large packed bed of Raflux rings 50-5

Tatyana S Petrova[1*], Daniela B Dzhonova-Atanasova[1], Krum A Semkov[1]

[1]Institute of Chemical Engineering, Bulgarian Academy of Sciences, Acad. G. Bonchev Str., Bl.103, 1113 Sofia, Bulgaria

* t.petrova@iche.bas.bg

Abstract. This work presents a continuation of our investigations on the radial liquid distribution in packed beds with open-structure random packings, by means of a dispersion model, in a scale, close to industrial one. Using experimental data for uniform initial irrigation, the optimal parameters' values of the three parameters of the dispersion model are obtained by two-parameter identification. One of the parameters (the radial spreading coefficient) is calculated independently by using experimental data for a point source initial irrigation. The dispersion model solution at optimal parameters' values is compared with both experimental and TUM WelChemCell model literature data for the liquid radial distribution in a packed column with a diameter of 1.2 m and random packing Raflux rings 50-5. The maldistribution factor of the liquid distribution is also calculated and compared. The comparison shows very good agreement between our results and the literature data and confirms the dispersion model capability to predict the liquid distribution in large columns with open-structure packings.

1. Introduction

In recent years, an increasing interest is observed in the investigation of the liquid phase maldistribution in packed columns with a large diameter [1-7], as well as in the application of open-structure random packings. These packings are characterized by a complex shape (a type of a 3D lattice, consisting of curved thin lamellae), low pressure drop, and high effective surface, which make them a worthy alternative to the structured packings, most often used in industrial scale transfer processes. Many of these packings are not investigate thoroughly, especially their radial distribution ability in columns with a large diameter. A series of our previous works is concerned with prediction of liquid distribution in metal random packings Raschig Super-Rings (0.7, 1 and 3") and metal Pall rings [4, 5, 7] by a dispersion model, but in columns with a diameter up to 0.6m. Our latest paper [6] was the first attempt to use the dispersion model for prediction of the liquid distribution at a large column diameter of 1.2m and for the metal RVT saddle rings 70-5, using the experimental data of [2]. The result encouraged us to continue and compare our results with recently published experimental and TUM WelChemCell Model results [3] for Raflux rings 50-5, obtained for the same column.

2. Problem formulation

As already mentioned, the experimental data for radial liquid distribution in a column with a diameter of 1.2m and a random metal open-structure packing Raflux-rings 50-5 (Figure 1) will be used in this study. The data are obtained at uniform initial distribution, liquid load L_0=20 $m^3/(m^2h)$, gas load factor F_g=1.3 $Pa^{0.5}$, bed height H=3m, and measured with a Liquid Collecting Device (LCD) [3]. This data is used to verify the capability of the dispersion model to predict the liquid distribution in an industrial scale. The LCD consists of 7 concentric annuli, each of them (excluding the 1st one) is divided into 3 equal 120° segments, i.e. the LCD divides the column cross-section into a total of 19 collecting segments [1]. Annuli area ratios R1÷R7 in % are (2.67, 8.00, 13.34, 18.67, 24.01, 29.35, 3.96). An extension of the TUM WelChemCell model is proposed in [3] and validated against experimental data for Raflux rings 50-5, showing good agreement for both uniform and point source initial liquid distribution.

Figure 1. Photography of a random metal packing Raflux ring 50-5.

As already stated in our previous works [4, 6, 7], the dispersion model solution needs identification of model parameters *B, C* and *D* for the respective packing and the current packing layer height $z = DH/R^2$, as can be seen in [6]. The solutions of the dispersion model for the dimensionless liquid density of irrigation, $f(r,z) = L_i/L_0$ and for the wall flow W^u, at uniform initial liquid distribution (superscript "u"), $r = r'/R$, have the form:

$$f^u(r,z) = A_0 + \sum_{n=1}^{\infty} A_n^u J_0(q_n r)\exp\left(-q_n^2 z\right), \quad W^u = \frac{1}{1+C} - 2\sum_{n=1}^{\infty} A_n^u(q_n)\frac{J_1(q_n)}{q_n}\exp\left(-q_n^2 z\right)$$
$$A_0 = \frac{C}{1+C}, \quad A_n^u = 2\left(q_n^2/B - 2C\right)\left\{\left[\left(q_n^2/B - 2C\right)^2 + q_n^2 + 4C\right]J_0(q_n)\right\}^{-1} \tag{1}$$

Following the procedure, presented in detail in [6], two independent approaches and two sets of experimental data (at uniform and point source initial irrigation) are used to identify the packing radial spreading coefficient *D, m*. Other two parameters are determined too, and comparison of dispersion model solution with experimental data and TUM WelChemCell Model results [3] is presented in the next section.

3. Results and discussion

The first approach to determine the packing radial spreading coefficient *D*, m is the so called single-jet method [8], which requires experimental data for point source initial irrigation. It is explained in detail in [8] and applied in [6] to an example with a random RMSR 70-5 packing, the same LCD and the same column diameter as those considered in [1]. For brevity, here only the final results for radial spreading coefficient *D* of Raflux rings 50-5 are presented in Figure 2.

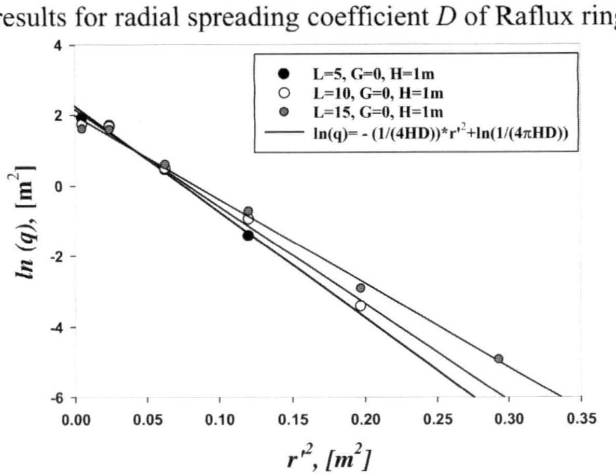

Figure 2. Values of *ln(q)* versus $(r')^2$ according to the single-jet method for determination of radial spreading coefficient *D* for Raflux rings 50-5, *H*=1m, point source initial irrigation data [3].

IOP Conf. Series: Materials Science and Engineering **876** (2020) 012009 doi:10.1088/1757-899X/876/1/012009

It was found that the mean average value of this coefficient is D=0.00939 m at 3 different liquid loads – 5, 10 and 15 m^3/(m^2.h), at G=0 and H=1 m. It is seen that, in contrast to the metal RVT saddle rings 70-5 [6], here the coefficient of radial spreading for Raflux rings 50-5 slightly increases with increasing of the liquid load. Practically, the Raflux rings have a quite different structure, which resembles that of the Pall rings. Compared to them, the structure of RVT saddle rings 70-5 is more open to the liquid flow.

The second approach to find D independently is the two-parameter identification procedure, minimizing the residual variance between model and experimental mean densities of irrigation in all 7 segments of the LCD. This method needs experimental data for radial liquid distribution at uniform initial irrigation, and the value of the parameter C (see Equation (1)). The parameter B is a criterion for exchange of liquid between the column wall and the packing; the parameter C expresses the equilibrium distribution of entire liquid flow between the wall and the packing when equilibrium state is attained $z \to \infty$ .The latter can be determined from experimental data of the wall flow at uniform initial irrigation and at least 3 different packing heights. In the dispersion model definition it is assumed that at some packing height the wall flow growth stabilizes (reaching a plateau), which corresponds mathematically to the case $z \to \infty$, and the formula $C = \left(1 - W^u_{z\to\infty}\right)/W^u_{z\to\infty}$, developed in [7], can be used. Here, the value C=7.12 was calculated using the experimental data for the wall flow in a bed of Raflux rings 50-5 at uniform initial irrigation, L_0=20 m^3/(m^2.h), F_g=1.3Pa$^{0.5}$, and packing heights 1, 1.5, 2.5 and 3 m.

The two-parameter identification of B and D are performed, and the results are given in Figure 3, as a 3D plot (a) and as a contour plot (b) of residual variance versus different values of B and D. The minimum of residual variance S_A^2 between model and experimental density of irrigation in each segment of LCD is obtained as 0.198e-01 for the optimal values of the parameters B=1.15, D=0.01m and C=7.12. As is seen, the obtained value for D from two-parameter identification (0.01m) is close to this, determined by single jet method earlier (0.00939m). For these optimal values of the parameters, the solution of the dispersion model as well as that for the wall flow (Equation (1)), were obtained. The results are compared with data for radial liquid distribution from physical experiment and calculations with TUM WelChem Cell model for Raflux rings 50-5 (both from [3]).

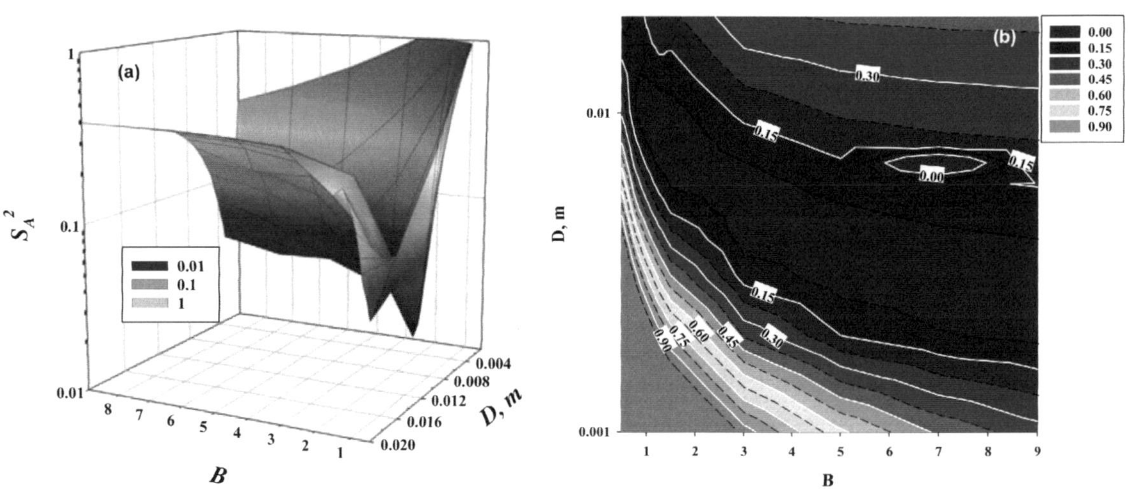

Figure 3. Visualization of two–parameter identification of model parameters with the minimum of residual variance as a criterion S_A^2=0.198e-01, at D=0.01 m, B=1.15, C=7.12, H=3m, for metal Raflux rings 50-5; (a) – 3D plot, (b) – contour plot.

The comparison is presented in Figure 4. As a whole, both models are in very good agreement with the experimental data. The relative errors in %, in segments R1÷R7 between the dispersion model predictions and the experimental data are 0.3;-2.4;-14.1;4.3;-10.2;11.6;-0.8, respectively. The dispersion model wall flow fits better to the experimental data (-0.8%), the predictions have accuracy of -14.1%/11.6%, which is quite reasonable for this column scale.

In [3, at figs.11-13] a detailed explanation about the deviations was given for the whole experimental database used (4 type of random packings, 4 packing layer heights, a wide range of liquid loads and gas load factors). It was noted there, on the base of an analysis of the parity plots for both of their models (original WelChem cell model and enhanced TUM WelChem Cell model) [3, see figs.12-13], that the original one does not take into account the liquid load and countercurrent gas flow effects, and moreover a simplified assumption for the wall flow is considered. The enhanced TUM WelChem Cell Model accounts for these effects and the wall flow assumption was refined by increasing packing porosity at the column wall.

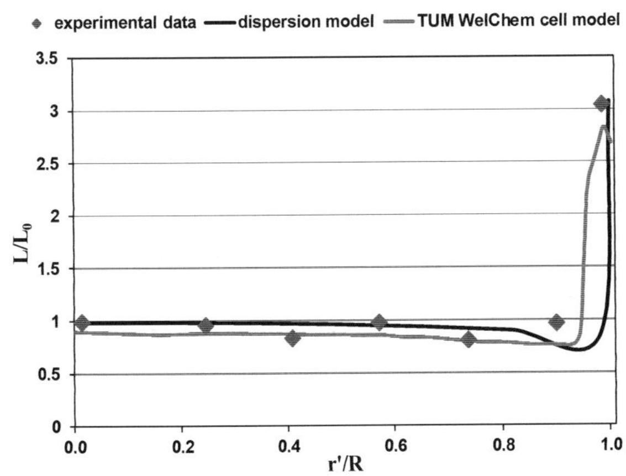

Figure 4. Comparison between experimental data for radial liquid distribution (blue rhombs), TUM WelChemCell model (red line) (both from [3, fig.11]) and dispersion model predictions (black line), for random metal Raflux rings 50-5, H=3m, column diameter 1.2m, liquid load L_0=20 $m^3/(m^2h)$, gas load factor F_g=1.3 $Pa^{0.5}$

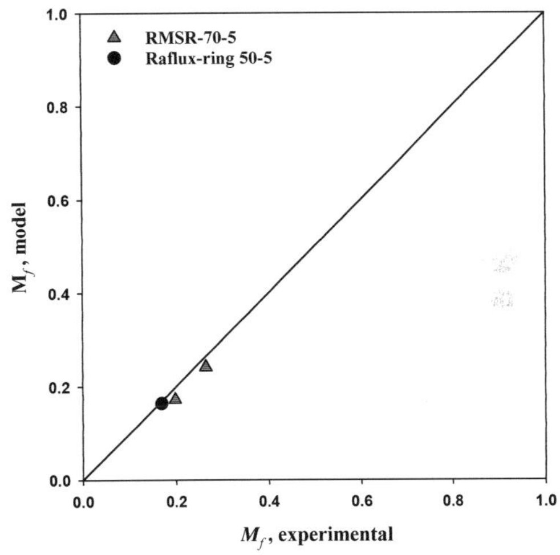

Figure 5. Comparison between experimental and model liquid maldistribution factors, calculated for Raflux rings 50-5 (this work) and for RMSR 70-5 [6].

The maldistribution factors of the experimental and model liquid distributions are calculated by the following formula [1,5]:

$$M_f = \frac{1}{F_0}\sum_{i=1}^{ns} F_i \left| \frac{L_i - L_0}{L_0} \right| \tag{2}$$

In equation (2) the ratio F_i/F_0 represents the ratio of the area of the respective segment i to the entire area of the LCD (i.e. column cross-section), m^2, ns is the number of segments of the LCD, L_i/L_0 is the dimensionless theoretical or experimental irrigation density in the segment i of LCD, limited between the radii r_{i-1} and $r_i\left(r_i > r_{i-1}\right)$.

In Figure 5 the maldistribution factors for the random packing Raflux rings 50-5, H=3m, are calculated and they almost completely coincide; the relative error is 2.5%. Also, our latest calculation results for the maldistribution factors are added based on the results in [6] for a random packing RMSR 70-5, and a packing bed height of 1 m and 2.5 m. For this packing the calculated values are M_f

= 0.172 for H=1m, and M_f =0.242 for H=2.5m. The corresponding experimental ones are M_f =0.198 for H=1m and M_f =0.265 for H=2.5m. The comparison between model results for liquid distribution and the respective experimental data at two packing heights shows good agreement - the relative errors are 13.1% and 8.6 % at H=1m and 2.5 m, respectively. The comparison shows good agreement between predicted and measured maldistribution factors for both packings in industrial scale column size.

4. Conclusions

In this work a modeling of radial liquid distribution in packed column with a diameter of 1.2 m and random metal packings Raflux rings 50-5 by the dispersion model is performed. Identification of the model parameters and comparison of obtained results with both experimental and TUM WelChem Cell model data [3], are presented. It was observed, that our predictions showed very good agreement with the results of [3]. The relative errors in %, in the segments of the LCD between the dispersion model predictions and the experimental data do not exceed 15 %, which is quite reasonable at this column scale. The relative error between the model and experimental maldistribution factors for Raflux rings 50-5 is 2.52%, and for RMSR 70-5, 13.1% and 8.6 % at H=1m and 2.5 m, respectively. The obtained results confirm the dispersion model capability to predict the liquid distribution and maldistribution in large columns with open-structure packings.

5. References

[1] Hanusch F, Rehfeldt S and Klein H 2018a Liquid maldistribution in random packed columns: experimental investigation of influencing factors *Chem. Eng. Technol.* **41**(11) 2241-2249.
[2] Hanusch F, Engel V, Kender R, Rehfeldt S and Klein H 2018b Development and application of theTUM-WelChemcell model for prediction of liquid distribution in random packed columns *Chemical Engineering Transactions* **69** 739 - 744.
[3] Hanusch F, Kender R, Engel V, Rehfeldt S and Klein H 2019 TUM–WelChem cell model for the prediction of liquid distribution in random packed columns *AIChE J.* **65** e16598, https://doi:10.1002/aic.16598
[4] Petrova T, Semkov Kr and Dzhonova-Atanasova D 2018 Modeling of liquid distribution in a packed column with open-structure random packings *Chemical Engineering Transactions* **70** 1051-1056.
[5] Dzhonova-Atanasova D, Petrova T, Semkov Kr, Darakchiev S, Stefanova K, Nakov S and Popov R 2018 Experimental investigation of liquid distribution in open-structure random packings as a basis for model refinement *Chemical Engineering Transactions* **70** 2077- 2082.
[6] Petrova T and Dzhonova-Atanasova D 2019 Flow simulation and identification of important model parameters in industrial packed beds for high-performance random packings *Journal of Ecological Engineering* 20 (9) 155–159.
[7] Petrova T, Stefanova K, Dzhonova-Atanasova D and Semkov Kr 2017 Modeling of a liquid spreading in randomly packed metal Pall rings *Mathematical Modeling* **1**(4), 174-177.
[8] Dzhonova-Atanasova D, Kolev N and Nakov Sv 2007 Determination of liquid radial spreading coefficients of some highly effective packings *Chem. Eng. Technol.* **30**(2), 202-207.

Acknowledgments

This work was financially supported by the National Science Fund at the Bulgarian Ministry of Education and Science, Contract No DN 07/14/15.12.2016.

AUTHOR INDEX

Institute of Physics
Dirac House, Temple Back
Bristol BS1 6BE UK

ISSN: 1757-8981
ISBN 978-1-7138-2689-7